RANCHO SANTIAGO COLLEGE
OR MAIN QB36.D68 W43 1983
Tree rings and telescopes : the
Webb, George Ernest.
3 3065 00031 2581
D0787201

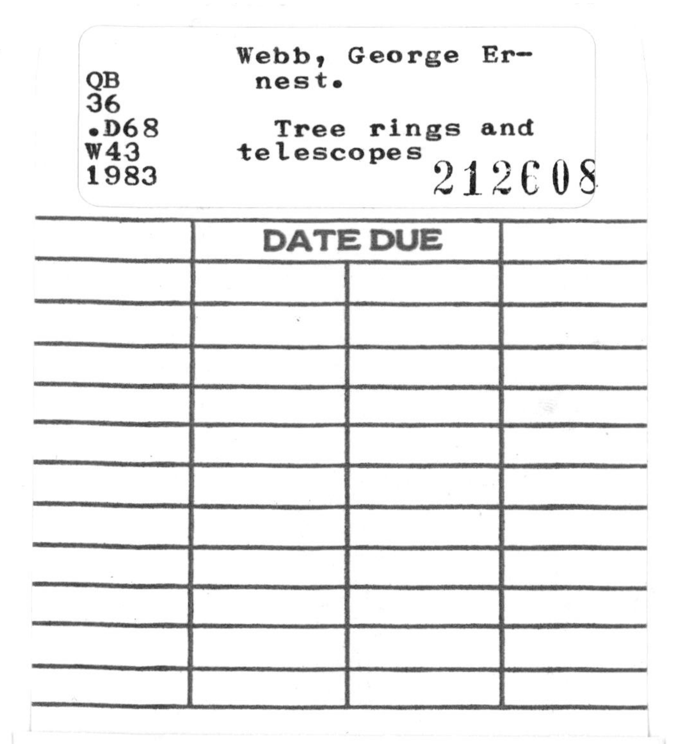
Webb, George Ernest.
QB 36 .D68 W43 1983
Tree rings and telescopes
212608
DATE DUE

Rancho Santiago College
Orange Campus Library

TREE RINGS AND TELESCOPES

TREE RINGS AND TELESCOPES

The Scientific Career of A. E. Douglass

GEORGE ERNEST WEBB

The University of Arizona Press
Tucson, Arizona

212608
Rancho Santiago College
Orange Campus Library
9-6-85

About the Author. . .

GEORGE ERNEST WEBB began his research in the work of A. E. Douglass in 1976 while he was studying toward his doctorate in history at the University of Arizona, which he received in 1978. A historian of science, Webb has published a number of articles in such journals as *The Astronomy Quarterly, Journal of American Culture, Arizona and the West,* and the *Journal of Arizona History.* In 1978 Webb joined the faculty of Tennessee Technological University, where he has given courses in the history of science and scientific controversies.

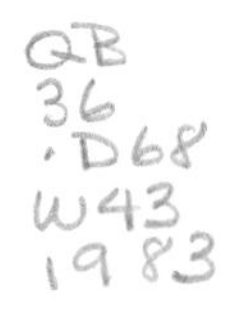

THE UNIVERSITY OF ARIZONA PRESS

Copyright © 1983
The Arizona Board of Regents
All Rights Reserved

This book was set in 11/12 V-I-P Aster
Manufactured in the U.S.A.

Unless otherwise noted, all photographs in this book appear courtesy of the Special Collections department of the University of Arizona Library.

Library of Congress Cataloging in Publication Data

Webb, George Ernest.
Tree rings and telescopes.

Bibliography: p.
Includes index.
1. Douglass, A. E. (Andrew Ellicott), 1867—1962.
2. Astronomers—United States—Biography. 3. Meteorologists—United States—Biography. I. Title.

QB36.D68W43 1983 520'.92'4 [B] 83-1152
ISBN 0-8165-0798-8

For E. J. W.
Semper adiuvans

Contents

Preface ix
1. THE YOUNG ASTRONOMER 1
2. THE CANALS OF MARS 20
3. CONFLICT AND ISOLATION 37
4. THE STEWARD OBSERVATORY 54
5. ASTRONOMY IN THE DESERT 69
6. A PUBLIC FIGURE 82
7. CYCLES, SUNSPOTS, AND PUEBLOS 101
8. THE SECRET OF THE SOUTHWEST 131
9. NEW CHALLENGES 152
10. THE FINAL QUEST 172
Notes to the Chapters 191
Selected Bibliography 226

Illustrations

A. E. Douglass at Trinity College, 1888 6
Harvard Observatory, Arequipa, Peru, 1893 11
Lowell Observatory, Flagstaff, Arizona, ca. 1901 34
A. E. Douglass and George Waterbury at Lowell Observatory 40
A. E. Douglass, ca. 1916 62
Steward Observatory Under Construction, 1921 72
Douglass Inside Steward Observatory Dome, 1922 75
Douglass at Steward Observatory, 1941 96
Multiple Plot of Sunspot Numbers, 1755–1911 112
Periodogram of Sunspot Numbers, 1755–1911 112
Douglass Near Forestdale Ruin, Arizona, 1928 120
Crossdating Samples 135
Beam HH-39 Being Excavated, 1929 142
Douglass in Tree-Ring Lab, 1940 162
Douglass in Tree-Ring Lab, 1941 174

Preface

On March 15, 1960, more than 100 persons gathered on the sun-warmed summit of a southern Arizona mountain to dedicate Kitt Peak National Observatory. From the beginning of the ceremonies, one man gained particular notice. Andrew Ellicott Douglass (1867–1962), ninety-three years old, listened intently to the speakers and carefully explored the site of the largest single collection of astronomical instruments in the United States. Throughout the day, old friends and colleagues made their way between telescope foundations and desert shrubs to greet him, while others asked the identity of the frail, white-haired gentleman. When they learned it, these younger men went out of their way to introduce themselves to the famous scientist. Educators, politicians, businessmen, and astronomers all wished to meet this living reminder of the advance of science in the American Southwest. The acknowledgment and esteem paid Douglass on this occasion was a fitting tribute to a long and distinguished career.

A. E. Douglass had lived in Arizona for more than six decades, ever since he joined the Lowell Observatory in 1894. Trained as an astronomer, he had already served with the Harvard College Observatory before Percival Lowell chose him as his principal assistant. Douglass later joined the faculty of the University of Arizona in Tucson in 1906, where he immediately mounted a campaign to bring a major astronomical observatory to southern Arizona, a region famous for its clear atmosphere.

Displaying the tenacity and dedication which characterized his entire career, Douglass continued his attempts for the next ten years, finally securing the needed funds in 1916. He guided the design and construction of the Steward Observatory and served as its first director until 1937, when he retired shortly before his seventieth birthday.

Retirement for Douglass merely meant the redirection of his energies to the continuation of his most famous work, the study of dendrochronology. Begun in 1904, this new science had already proved itself a valuable adjunct to archaeology and anthropology by providing a means for dating the ruins of the Southwest. Douglass's goal in the study of tree rings, however, had always been to find evidence of climatic cycles which could be used to predict terrestrial weather. This work consumed the remaining years of his life. Only his death in 1962 extinguished Douglass's dedication to the advance of scientific knowledge.

Although displaying great diversity, Douglass's work was united by several important themes. The unique environment of the American Southwest early impressed itself on the Vermont native when he surveyed the area for the Lowell Observatory in 1894. In addition to providing the clear atmosphere vital for astronomical research, the arid climate of the region also suggested to Douglass an important modification to the interpretation of tree growth. A few years after his arrival in Arizona, he recognized that the territory's trees responded chiefly to precipitation in their growth. This characteristic, less significantly displayed in other climates, resulted in a record of past rainfall variations displayed in terms of tree-ring widths. Douglass used the resulting patterns to establish the long chronologies which were eventually responsible for dating the pueblo ruins of the Southwest.

In all his scientific pursuits, Douglass displayed his dedication to the collection and analysis of data. In this, he differed from more famous scientists whose contributions grew from a singularly brilliant flash of interpretation which led to a new theory or a more accurate view of the universe. Douglass proved to be a extremely diligent researcher, but his contributions resulted as much from enthusiasm and perseverance as from the clear insight which characterizes the traditional scientific "genius." If he lacked the inspired imagination of

Einstein, however, he possessed in abundance the dogged determination of Darwin.

Further, while participating in the modern growth of American science, Douglass remained very much the solitary scientific investigator of earlier centuries. In a period of growing institutionalism, he was primarily an individualist. Although several large institutions assisted Douglass, his ideas and their application remained primarily those of a single man. Indeed, in dendrochronology, the field of study for which he would become best known, Douglass was the only active researcher for two decades, untill he began training others to carry on his work.

Douglass lived during one of the most dynamic periods in history of American science. When he began his study of astronomy in the 1880s, the United States displayed little interest in scientific research. As he continued his work in the early twentieth century, however, the nation's reputation began to grow through the endeavors of such men as George Ellery Hale in astronomy and Robert A. Millikan in physics. Douglass's dramatic investigations into the dating of archaeological ruins in the 1920s and the 1930s occurred simultaneously with Robert H. Goddard's pioneering work with liquid-fueled rockets. His later attempts at long-range weather prediction paralleled the advance of nuclear energy and the construction of the world's largest telescope on Mt. Palomar in California. During his final year of life, five astronauts penetrated the blackness of space.

For three quarters of a century, Douglass viewed his professional concerns as the dominant factor in his life. In accordance with Douglass's own evaluation of his significance, this examination of his career focuses on these scientific pursuits and the role they played in the advance of science in the American Southwest

Acknowledgments

Many debts have been incurred during the preparation of this volume. An examination of Douglass's career was first suggested by Harwood P. Hinton of the University of Arizona, whose comments and suggestions have been most helpful. David W. Corson (of the University of Arizona before going on to Cornell University), who first introduced me to the fascination of the history of science, early encouraged me to pursue Douglass's scientific odyssey and greatly aided in the correction of several serious lapses in early drafts. Paul A. Carter has been an inexhaustible source of valuable ideas over the past several years and suggested many improvements in the literary quality of the manuscript. I owe a particular debt to Herman E. Bateman, whose constant encouragement and friendship assisted greatly in the completion of this work. Bryant Bannister, former director of the Laboratory of Tree-Ring Research at Arizona, contributed many hours of his limited time to share his knowledge of Douglass and dendrochronology. A special note of thanks is due to Phyllis Ball of the Special Collections Department of the University of Arizona Library, who labored for several years arranging reams of notes and letters into the well-organized Douglass Papers, which were a pleasure to use. A final acknowledgment goes to Edith Douglass, who shared memories of her uncle and graciously encouraged the preparation of this study.

Many others at the University of Arizona have assisted in this work, particularly Roger L. Nichols, Richard R. Willey, William J. Robinson, Marilyn Bradian, Dorothy Donnelly, Dawn Polter, and Lynn Cresson. Marshall Townsend, Elizabeth Shaw, and Gregory McNamee of the University of Arizona Press contributed significantly to the production of this volume in a professional sense by providing valuable suggestions and a pleasant atmosphere. A number of colleagues at Tennessee Technological University have been most helpful, but I wish to express my special gratitude to B. F. Jones, Richard C. Lukas, Stuart O. Stumpf, William J. Brinker, and Lois Richardson for their valued support and friendship. To all those who gave encouragement and support, I acknowledge a special dept which can never be fully repaid.

I should also like to thank the following for permission to reprint documentary materials under their ownership: the California Institute of Technology Archives, for use of the George Ellery Hale Papers; the Harvard College Observatory, for use of the Pickering Papers and observatory records in the

Director's Files; the *Journal of Arizona History*, for use of my article, "The Indefatigable Astronomer: A. E. Douglass and the Founding of the Steward Observatory," (Summer, 1978) which appears here in altered form; the Mary Lea Shane Archives of the Lick Observatory, University of California, Santa Cruz, for use of Douglass letters in their holdings; the Lowell Observatory, Flagstaff, Arizona, for use of the Percival Lowell Papers, A. E. Douglass Papers, Letter Books, W. L. Putnam Correspondence, and Douglass's draft of "The Lick Review of 'Mars';" the Museum of Northern Arizona, Flagstaff, for use of museum archives and manuscripts; and the *Astronomy Quarterly*, in which parts of this book appeared in different form under the copyright of the Pachart Corporation, reprinted with permission of the director, Pachart Publishing House, 1130 San Lucas Circle, Tucson, Arizona 85704.

G.E.W.

1

The Young Astronomer

THE SUMMER OF 1867 was a comparatively peaceful time in rural New England. The horror of the Civil War had faded into a haunting memory, and the problems of Reconstruction seemed sufficiently distant to cause little concern. The return of peace brought with it a revitalization of the intellectual vigor which had long characterized the region. Traditionally the cultural center of the United States, New England entered the postwar era with expectations of significant intellectual achievement. On July 5, 1867, in Windsor, Vermont, a child was born into this stimulating, yet peaceful environment. His name was Andrew Ellicott Douglass.

The fifth of six children born to Reverend and Mrs. Malcolm Douglass, Andrew possessed a remarkable intellectual heritage. The first family member to gain scientific fame was his paternal great-grandfather and namesake, Andrew Ellicott (1754–1820), who was one of the most gifted surveyors of his day. Ellicott surveyed the boundaries of Pennsylvania (briefly working with the famous American astronomer David Rittenhouse), assisted Pierre L'Enfant in laying out the District of Columbia, and served as commissioner to establish the boundary between the United States and Spanish Florida. He then became secretary of the Pennsylvania Land Office, and was Professor of Mathematics at West Point until his death in 1820. Ellicott's daughter Ann Eliza married in 1815 Major David Bates Douglass (1790–1849), a colleague at West Point. A graduate of Yale, Douglass was a topographer and botanist on the Lewis Cass western expedition in 1820, served as president of Kenyon College, and then taught mathematics and natural philosophy at Hobart College in New York.[1]

His son Malcolm, Andrew's father, attended Trinity College in Hartford, Connecticut, and pursued graduate work at Hobart College, where he met Sarah Elizabeth Hale, daughter of the college president. After receiving master's degrees from both Hobart and Trinity, Malcolm Douglass completed his training as an Episcopalian minister and married Sarah Hale in 1851. A popular rector of several New England churches, Douglass established such an impressive reputation that his fellow clergy urged him to become bishop of their diocese in 1868. He modestly declined their invitation.[2]

A.E. Douglass grew up in comfortable Episcopalian rectories with a stimulating intellectual atmosphere. His great-grandfather's scientific instruments, including a small telescope, had passed to Douglass's family and proved of great interest to the young scholar. Recognizing Andrew's growing interest in astronomy, an aunt gave him a star atlas shortly before his eleventh birthday, a gift which the young boy put to immediate use with the Ellicott telescope. Andrew attended Central Grammar School in Andover, Massachusetts, proved himself a superior student, and because of his growing interest in science was asked to give a brief presentation on the telescope in the closing ceremonies on June 17, 1880.

At Punchard Free (High) School, Andrew expanded his scientific endeavors. At the end of his first year, he presented a paper before the school's Literary and Scientific Society discussing his lens-grinding experiments. The next fall, he observed a solar eclipse and published a school newspaper article describing the halos visible around the sun. Douglass's performance in high school sufficiently impressed his teachers that they allowed him to teach an astronomy course as a senior and to conduct the demonstration experiments for physics classes. Not surprisingly, he was one of the speakers at his graduation in June of 1884. Addressing the topic of education, Douglass emphasized that its true purpose was to develop the individual's talents, and criticized schools for failing to accomplish this purpose.[3]

Douglass's graduation remarks reflected his impatience with studying subjects of little use to a scientist. The results of his admission examination for Trinity College also mirrored his single-mindedness regarding education. He failed those subjects which he felt did little to develop his talents: Greek, Cicero, Roman History, and Ancient Geography. But he also failed to pass geometry, an essential component of any science curriculum. Douglass's college career would have to wait.

Because his father had been a Trinity graduate, Douglass's failure to secure admission on his first attempt was a cruel blow. He spent the next year diligently studying those subjects in which he had done badly, but also found time to observe a partial solar eclipse on March 16, 1885. The following June, Douglass again took Trinity's entrance examination. He was asked to discourse on Greek and Latin grammar, translation, composition, and history, and also to work problems in common arithmetic and elementary geometry. On June 26, the examiners informed Douglass that he had passed.[4]

Douglass spent four years at Trinity College. He won the class Geometry Prize in the second semester of his freshman year (partially redeeming himself for his first admission examination), and as a sophomore achieved an "A" grade in physics, botany, and zoology, as well as in religious instruction. The junior year proved to be an interruption in Douglass's enjoyable college career. In late September, 1887, his father Malcolm died of apoplexy in Wareham, Massachusetts, where he had been engaged in missionary activities. Deeply saddened by his loss, Douglass nonetheless persevered in his academic pursuits. His essay on "Aluminum" won Trinity's Chemical Prize for 1888 and he was elected to Phi Beta Kappa in the spring. During his final year at Trinity, Douglass maintained his superior grades and on June 27, 1889, received the degree of Bachelor of Arts with honors in astronomy, mathematics and physics.[5]

Although he focused his attention on scientific matters, Douglass allowed himself a few diversions while in college. As had his father, he joined the Beta Beta chapter of the social fraternity Psi Upsilon and was moderately active in that organization. Of greater intellectual interest were his modest investigations of the occult, which included attendance at a seance and at exhibitions on mesmerism. He also collected a body of notes concerning the Seybert Commission on Spiritualism, appointed by the University of Pennsylvania in accordance with the will of the wealthy spiritualist Henry Seybert. Douglass approached these events with a cautiously open mind, but searched for logical explanations of the supposed mystical phenomena.[6]

At Trinity, Douglass's primary interest centered on the college observatory. In an early letter to his parents, Douglass spoke enthusiastically of Flavel Sweeten Luther, Seabury Professor of Mathematics and Astronomy. Luther also served as director of the observatory, which housed a six-inch refracting telescope and accessories. The Trinity astronomer proved equally impressed with young Douglass, hiring him to help prepare a sunrise/sunset chart for the Hartford area. In addition to the salary of twenty-five dollars for this task, Douglass earned the respect of his mentor, who in December of 1885 gave Douglass a key to the observatory so that he might come and go as he wished. The young astronomer's career had begun.[7]

By the time of Douglass's formal alliance with his chosen discipline, the field of astronomy had become one of the most dramatic sciences of the era. Astronomers were greatly expanding their knowledge throughout the nineteenth century. Improvements in telescopes hastened the discovery of the asteroids between Mars and Jupiter, with more than 400 charted by the end of the century. Throughout the second half of the century, astronomers employed the recently invented spectroscope to examine the composition of the light emitted by stars and nebulae. The studies of the British astronomer Sir William Huggins in the early 1860s showed that many nebular spectra displayed a wide distribution of colors, suggesting that these nebulae were composed of myriad stars rather than clouds of gas. A few years later, Huggins spectroscopically determined that various stars and nebulae were approaching Earth while others seemed to be receding from the planet. Similar advances continued in all branches of astronomy during Douglass's early career.[8]

Douglass immersed himself in his observatory work, quickly taking advantage of a unique opportunity in the spring of 1886. The planet Mars, one of the most intriguing astronomical objects, made one of its periodic close approaches to Earth (known as an opposition) in March and April. When the red planet appeared, Douglass focused the Trinity telescope on Mars and sketched as many surface details as possible with the instrument's 180-power magnification. He studied the north polar cap and various dark surface markings he called "continents," which reflected the widely held opinion that Martian topography resembled that of Earth. Douglass later made observations of sunspots to determine the rotation period of the sun, arriving at the figure of twenty-five to twenty-seven days.[9] His calculations were close to nineteenth-century estimates, but proved some six days short of more accurate measurements in later years.

Douglass developed a keen interest in photography during his undergraduate years, an interest which complemented his endeavors in astronomy. During his last year at Trinity, he developed a number of plates showing astronomical and meteorological phenomena. These included photographs of the moon and of a solar eclipse, a photographic record of laboratory-produced electric sparks, and a dramatic exposure of a lightning storm near the Trinity campus. Douglass also adapted the recently developed technique of dry-plate photography to astronomy in his junior year. This technique proved of such value that Trinity added it to its astronomical curriculum. Douglass's endeavors at the Trinity observatory led Luther to characterize him later as "the best student I ever had."[10]

Toward the end of his studies at Trinity, Douglass gave considerable thought to his future. As did many other young Americans of his day, he longed for the opportunity to undertake graduate study in Europe, but an alternative for advanced education existed. Several American colleges had begun to offer graduate work, with Johns Hopkins University in Baltimore the leader in this development. Because of his father's death, European study would require a greater monetary investment than Douglass could afford, but Johns Hopkins remained a possibility throughout the summer following his graduation. Study at the Baltimore school also proved an insurmountable financial burden, however, which prevented Douglass from pursuing further education.[11]

Douglass faced a choice in employment following graduation. Through family and friends, his mother secured the promise of a teaching job in Maine. The young astronomer,

A. E. Douglass at Trinity College, 1888.

however, already had promising hopes for a research position. He had written earlier to Edward C. Pickering, director of the Harvard College Observatory, to borrow a catalogue outlining graduate work in astronomy. When he returned the pamphlet, he wrote Pickering that he planned to visit Boston in early September in hope of finding employment in the observatory.

Neither Pickering nor his younger brother William (a distinguished astronomer in his own right) was there when the young man visited, so Douglass wrote to the director and formally applied for a position. Douglass wanted to spend one or two years at the observatory, and take a few graduate courses in his spare time.

Douglass applied at precisely the right moment. William H. Pickering had been searching for an assistant to work mornings and afternoons throughout the week, and invited him for an interview. Douglass sufficiently impressed the younger Pickering to secure an assistantship with the observatory at an annual salary of five hundred dollars.

Douglass spent his first year at Harvard familiarizing himself with the observatory and assisting the Pickerings in their research. As a junior assistant, Douglass's main duty was the recording of data from the many continuing projects, but he also spent considerable time at the observatory's telescopes, improving his technique and adding to his astronomical experience. In mid-1890 a valuable opportunity presented itself to Douglass. W. H. Pickering chose him as one of the assistants for his expedition to establish a branch observatory in the Peruvian Andes. Designed to take advantage of the superior geographical location and altitude (which would enable the study of the relatively unknown southern skies), Harvard's Andean station would be Douglass's home for the next three years.[12]

Harvard had been searching for such a site since the spring of 1887, to escape the atmospheric impediments already plaguing astronomy. Uriah A. Boyden (1804–1879), a successful Boston mechanical engineer, had willed $238,000 to support the establishment of a high-altitude observatory, a fund transferred to Harvard in early 1887. During the summer, the Pickering brothers examined sites in the Colorado Springs and Pike's Peak areas of Colorado, but found atmospheric conditions little better than in Cambridge. During the winter of 1888–89, W.H. Pickering and his party installed a telescope on Mount Wilson, east of Los Angeles, at an altitude of 6000 feet. The Harvard group remained on this mountain more than a year, finding atmospheric conditions far superior to those in Colorado, but could not gain clear title to the land.[13]

The Arequipa Observatory

Simultaneously with the California expedition, another Harvard group conducted a survey along the west coast of South America. The leader of the party, Solon I. Bailey (later to serve as director of the Harvard Observatory), located a desirable station near Lima, Peru. The team spent a year testing this site, but Bailey sought a location nearer the tropics. Examining the coastal area south to Valparaíso, Chile, the Harvard astronomers recommended that a station be constructed at Arequipa, Peru, at an altitude of 8100 feet. After carefully reviewing Bailey's reports, the Harvard staff organized an expedition to establish the Boyden Station of Harvard College Observatory at Arequipa. The station would be directed by W.H. Pickering and his principal assistant, A.E. Douglass.[14]

The Harvard expedition left New York for South America on the S.S. *Colón* the afternoon of December 20, 1890. Reaching Colón, Panama, on the twenty-eighth, Pickering, Douglass and the rest of the expedition spent the next three days visiting points of interest in the area. They sailed on the S. S. *Serena* on New Year's Day, and reached Arequipa on January 17, 1891. Douglass and George T. Vickers, a fellow assistant, had spent part of their voyage studying Spanish with a Señor Baguer, the Spanish consul to Callao, Peru, who happened to be on board. Both young men proved able pupils and could quickly converse with the natives of the area.[15]

The personnel of the Boyden Station examined sites in the Arequipa region and made plans for the construction of the observatory. They obtained living quarters in a house owned by Mr. MacCord, the director of the local railroad, and then moved quickly to select a site two miles northwest of the city and five hundred feet above it. Pickering purchased approximately two acres of land there and in February of 1891 began laying the foundation for the observatory. Throughout the spring and summer, Douglass and his colleagues labored to construct buildings and install and adjust the astronomical instruments. As fall approached, the three telescopes and laboratory were operating and the observers' house was nearing completion. Scientific work at Boyden Station could now begin.[16]

In addition to capitalizing on the exceptionally clear skies of the Peruvian Andes, Harvard's southern observatory would expand important astronomical work to the southern hemisphere. The value of observations south of the equator had long been recognized but, with the exception of the British astronomer John Herschel's studies of the southern skies from the Cape of Good Hope in 1833–38, little serious work had been accomplished. As director of the Harvard Observatory, E.C. Pickering desired data from the southern skies to complement the observatory's study of variable stars, novae, star clusters, and nebulae. He also realized that the celestial photography library recently begun in Cambridge would be of much greater value if it included photographs of the southern skies.

As soon as the telescopes were in operation, therefore, Douglass initiated the extension of earlier work to the southern hemisphere. His first task was a search for double stars, in an attempt to locate possible binary star systems (in which two stars revolve about each other). While successfully locating several doubles, he also detected a few nebulae, cloud-like phenomena which had recently been observed through large telescopes to present spiral shapes. The location and description of these nebulae represented data of great value, which would enable astronomers to judge whether these bodies were external galaxies similar to the Milky Way. Douglass also assisted in the photography of southern stars, beginning an eight-year program at Arequipa which ultimately recorded almost 6000 photographs of the spectra of more than 1100 stars. These plates were placed in the charge of Harvard's Annie J. Cannon, who sorted and organized the observatory's collection of star photographs into spectral classes.[17]

While scanning the Andean skies for stellar and galactic phenomena, the Arequipa astronomers also conducted significant planetary research. They early turned their instruments toward Mercury, Mars, Venus, Jupiter, and Neptune, taking advantage of the clear Arequipa atmosphere and the favorable positions of the planets. Their observations of Mercury during July and August of 1892 disclosed surface markings which, when traced over a long period, seemed to confirm the belief that the small planet presented the same face toward the sun at all times. This theory held sway until the mid-1960s, when radar studies showed that Mercury's rotation took less time than its revolution. The Harvard astronomers also calculated the diameter of Venus at 7662 miles (very close to later estimates), and correctly concluded that the density of the Venusian atmosphere greatly exceeded that of Earth.[18]

Douglass and his fellow astronomers also scrutinized the outer planets of the solar system, particularly Jupiter and Neptune. Because the dense atmosphere of Jupiter precluded surface observations, they studied its four largest satellites. Discovered by Galileo in 1610, these moons seemed to show surface bands or belts. Pickering and Douglass calculated the mean diameters of the satellites, significantly overestimating the diameters of the first three, but arriving at an accurate estimate of the size of Callisto. In studying Neptune, Douglass recorded bands or belts on the planet, and accurately calculated the diameter of the body.[19]

The personnel of the Boyden Station also investigated Mars. During the preceding decade, many of the world's astronomers had taken advantage of each opposition in an attempt to confirm the 1877 discovery made by Giovanni Schiaparelli (1835–1910). Using a medium-sized refractor in Milan, Italy, Schiaparelli had noted a number of dark, straight lines on the planet's surface, which he called *canali,* a general Italian term for "channels" or "grooves." The translation of his work into English, however, resulted in the use of the word "canals" to describe these surface markings, suggesting engineering activity on the red planet.[20] During the 1880s and 1890s, these dark streaks remained a major research topic in planetary astronomy.

Douglass and Pickering devoted much of their time to Mars during the opposition of summer 1892. Cloud belts and canals remained clearly visible, confirming the research of Schiaparelli and others. In addition, at several of the intersections of canals, they noted small spots which Pickering called "lakes" for convenience. The most significant discovery, however, occurred on June 24, 1892, when the astronomers observed streaks similar to Schiaparelli's canals in the dark areas of Mars. Confirmed by observations in mid-July, these streaks challenged the theory that the dark markings represented large bodies of surface water. To check their results,

Harvard College Observatory, Arequipa, Peru, in early 1893.

the astronomers erected a disk eight feet in diameter some eleven miles from the observatory on the flank of Mount Chachani. Drawing various lines and patterns on this device, they attempted to determine if the markings on Mars could be the result of optical illusions or telescope imperfections. This research, however, proved inconclusive.[21]

Douglass also became involved in a number of other scientific enterprises. As time permitted, he took geological notes on trips in the area and made photographic records of the zodiacal light. He considered the latter his most valuable personal contribution to the expedition because no careful study of the phenomenon had been attempted in the southern hemisphere. He also collected information on the earthquakes of the area, but even after experiencing a violent tremor, he remained anxious for further personal knowledge. After a minute-long quake in February of 1891, he wrote his sister: "The shake is said to be a violent one, but if we dont [*sic*] get any better ones than that, I shall be disappointed."[22]

When not working at the observatory, Douglass made several interesting trips, journeying throughout Peru and to Bolivia, and climbing several mountains over 16,000 feet high. He developed an interest in archaeological ruins and began to collect artifacts and fossils. This brought him to the attention of F.W. Putnam, the curator of Harvard's Peabody Museum of American Anthropology and Ethnology. Putnam planned to send an assistant to Peru to collect specimens for the Peabody exhibition at the 1893 Chicago World's Fair and asked Douglass to show the man around, a task the astronomer eagerly accepted.[23]

As the spring of 1893 approached, the Harvard expedition prepared to leave the new observatory to their replacements and return home. Once again, Douglass considered his future. He toyed with the idea of graduate study at Johns Hopkins, but decided that keeping his post with the Harvard Observatory would be professionally and economically wise. He planned to remain at Cambridge and take classes at Harvard, if possible.[24]

The Harvard party left Arequipa on March 27 and boarded the S. S. *Itata* the next day. On April 16 they landed at Vallenar, Chile, and set up a temporary camp to observe an impending solar eclipse. A week later, they photographically recorded the event. Continuing southward, the group passed through the Straits of Magellan on May 13, and continued the organization and classification of their recent data.

As had been planned, Douglass and Pickering remained on board after arriving at Rio de Janeiro and continued their journey to Europe to visit several observatories. Reaching London on June 24, they went to the Greenwich Observatory and subsequently crossed the channel to inspect the French national observatory in Paris. After this, Douglass secured a month's extension on his vacation and visited Stonehenge, Southampton, and Brighton. On July 27 he sailed from Liverpool for Boston, and upon reaching home spent ten days with his family.

Douglass and Pickering left Boston on August 19 for the World's Fair and the Astronomical Congress being held in Chicago at the same time. Douglass divided his time between the two events, attending the astronomical meetings and visiting the exhibitions at the fair. Beginning on September 1 he resumed his work for the Harvard Observatory, further reducing the data from South America. Then, in January of 1894, he met Percival Lowell, the wealthy amateur astronomer who would play a major role in the next phase of his career.[25]

Douglass and Lowell

Percival Lowell had already enjoyed an interesting life before he entered astronomy. Born on March 13, 1855, Lowell had descended from the Lowells and Lawrences, two prominent Boston families. Educated in Boston preparatory and French boarding schools, he entered Harvard in the fall of 1872, gaining election to Phi Beta Kappa in his senior year and giving one of the commencement addresses. While building his own fortune through family business interests, Lowell retained his interest in astronomy which had begun in adolescence. He observed Mars, read widely on astronomical topics, and honed his mathematical skills. On a trip to Japan in 1892, Lowell carried a medium-sized telescope with which he observed Saturn.[26]

In 1892 Lowell learned that Schiaparelli, whom he greatly admired, had abandoned his work on the Martian *canali* because of failing eyesight. Hoping to continue the Italian astronomer's work, Lowell eagerly anticipated the next opposition of Mars in the summer of 1894. He enthusiastically responded to W. H. Pickering's suggestion that an observatory be built to continue the astronomical investigation of Mars. He agreed to finance the observatory and asked Pickering to join his staff. Pickering in turn recommended Douglass as an assistant and arranged leaves of absence for the two Harvard astronomers.[27]

Because of the limited time before the next opposition, Lowell decided to send Douglass immediately to test possible sites in the Far West. Pickering recommended that Douglass check the deserts of the Southwest, especially those of southern Arizona. Lowell accepted the advice, and instructed his assistant to observe two nights each in Tombstone, Tucson, and Phoenix, the major population centers in the territory. On February 28, 1894, Douglass boarded a train in Boston for the trip west, carrying with him Lowell's telescope.[28] During the week-long journey to Arizona, he peered through the train window at the gegenschein, the faint glow in the sky opposite the setting sun, and made careful notes of his observations. Douglass arrived at the Arizona town of Benson after dark on March 7.[29]

Douglass's primary task was to find an observatory site with the best combination of clear and steady air. The large number of cloudless days in southern Arizona suggested that clear air would be prevalent, but finding steady air was another matter. Douglass thus had to gauge carefully the existence and extent of damaging atmospheric currents. He observed these currents by removing the eyepiece from the telescope and placing his eye at the focus. In this way, wave-like appearances could be seen crossing the field of view when the instrument pointed toward a bright star or small planet. W.H. Pickering had devised a scale of "seeing" based on the spurious rings formed around the light source as viewed through the telescope with its eyepiece removed. This first attempt to establish a scientific method for evaluating observatory sites would guide

Douglass in his estimates. The scale defined ten as the maximum value:

10. "Star distinct from inner ring. Rings motionless."
8. "Star distinct from inner ring. Rings complete but moving."
6. "Star more or less confounded with inner ring. Rings broken into dots and lines but still traceable."
4. "Occasional evidence of rings."
2. "Image fairly small but in motion. No evidence of rings. Appearances like that usually seen in Cambridge when not very bad."
0. "Image larger and in constant motion. Constant change of size. Useless for observation."[30]

On March 8 Douglass, carrying Pickering's scale and Lowell's telescope, boarded the stagecoach at Benson for Tombstone shortly after dawn. Three hours later he arrived at the famous mining town and contacted A. H. Emanuel, clerk of the District Court, who provided him with a free room. Douglass quickly set up his telescope in Emanuel's yard. The Tombstone townspeople proved curious, so he spent his first ninety minutes in town showing his audience various objects through the instrument. That night, Douglass began his observations in the Tombstone area. At 9:00 P.M., he sighted Jupiter and found the seeing about ten. An hour later, however, a slight breeze picked up, lowering the seeing on several bright stars to six or lower for the rest of the night.[31]

In a letter to Lowell written the next day, Douglass commented on the Tombstone area, sending the reassuring news that the town "is claimed to be perfectly safe either from desperadoes or Indians." In addition to being safe from violent elements, he continued, Tombstone also presented itself as a potentially helpful community. George H. Fritz, mayor of the city, told Douglass that the town would donate any land selected for the observatory and would also build a road if necessary. A. H. Emanuel offered part of his property as an observatory site and agreed to provide free lodging for astronomers, who would be responsible only for cleaning expenses and meals.[32] While perhaps not realizing the scientific value of an observatory, Tombstone residents appreciated the publicity value of such an installation.

Douglass searched for more promising observation points near Tombstone. Seeking a fairly high spot, he found an encouraging and accessible site on the tailings dump of the Grand Central Mine, approximately 250 feet above the level of the town. A third site was on one of the nearest of the Tombstone Hills, at an altitude of 4770 feet, almost 400 feet higher than the first site tested. The seeing at each spot proved good, but certainly not excellent, usually giving values of six to eight. The wind seemed to be the chief culprit in causing the air's instability. After two days of tests, Douglass left Tombstone shortly after noon on March 11. Reaching Benson, he boarded a freight train and arrived in Tucson after dark.[33]

From the first, Douglass was wary of the Tucson atmosphere. As he stepped off the train, he noted that the stars twinkled very badly, an indication of unsteady air. Also, he soon learned that the Tucson area had not yet recovered from a severe winter storm which had dropped a foot of snow on Oracle, a village northeast of the city. Nonetheless, Douglass began observations in Tucson on March 12. Renting transportation, he took his telescope to a hill 250 feet high located south of town near a stone quarry. This was the highest hill readily accessible, but it proved unsatisfactory. The seeing there measured four, and although it occasionally reached eight, by 9:30 P.M. a cold south wind dropped the rating to two. Douglass abandoned the hills to the south, and looked to the low mountains west of town.[34]

Douglass next set up his telescope on a sheltered site between the summits of Sentinel Peak and Turtleback (now Tumamoc) Hill. Seeing values ranged from six to ten, but again by 9:30 a slight haze covered the sky, dropping the seeing to four. Although the weather worsened over the next few days, the young astronomer investigated the summits of Sentinel Peak and Turtleback Hill, finding the latter promising. Shortly before leaving Tucson for Phoenix, Douglass sent Lowell the results of his observations. The Turtleback Hill site appeared the best to Douglass, although the solid rock of the summit would require blasting to install the telescope. He ranked the two sites outside Tombstone next and rejected the other three.[35]

Of all the Arizona towns, Tucson most wanted the observatory, but the city's attempts had begun inauspiciously. Theodore B. Comstock, president of the University of Arizona and the community's intellectual leader, was out of town when Douglass arrived and no one else had come to meet the train. After this lapse, however, the town's campaign began quickly. A few days after Douglass's arrival, the editor of the *Weekly Citizen* urged that Tucson should do everything possible to get the observatory. "[N]either courtesies nor money should prevent us from securing it if possible," the editor commented. Tucsonans continued their attempt to secure the facility throughout Douglass's Arizona travels, assuring him that the weather during his visit was quite unusual and emphasizing Tucson's willingness to donate land and labor as needed. The astronomer's sights, however, had been diverted to the north. The uncharacteristically bad atmosphere led Douglass to abandon the area which seventy years later became the center of visual astronomy in the United States.[36]

Douglass boarded a train in Tucson near midnight on March 16, and arrived in Phoenix shortly after six the next morning. The first night in Phoenix caused him great concern. Clouds, a fairly strong wind, and the lack of an adequate hill all combined to raise serious doubts about the valley. On the recommendation of W.J. Van Horn, President of the Tempe Board of Trade, Douglass took his telescope a few miles east to the Tempe Buttes and set up his instrument on a 260-foot-high hill. The results were discouraging: a forty-percent cloud cover and a heavy wind made observations difficult. Even when the wind died down later in the evening, seeing proved no better than six.[37]

Douglass began to doubt the ultimate success of his mission. He telegraphed Lowell, asking whether he should wait for satisfactory weather or abandon southern Arizona to examine Prescott (140 miles to the north), which had been recommended by several persons in Tucson. In a letter written the same day, Douglass told his employer that he thought it would be worthwhile to wait for good weather in Tempe. If the tests proved exceptional, he might check on Yuma, in the extreme southwest corner of the state, whose climate was similar to Tempe's. Either way, in Douglass's view, the Tempe observations would determine the course of the rest of the survey. Lowell cabled instructions on March 20 to wait for good weather in Tempe and then try Prescott if it still looked promising.[38]

Three days later, the weather cleared, and Douglass resumed his observations. On the nights of March 23 and 24, Douglass encountered patterns similar to most of his recent tests. The seeing was at first fairly promising, but as soon as the wind increased the seeing dropped. Douglass telegraphed the latest results to Lowell, who advised him to examine Prescott and then to test Flagstaff, Arizona, and perhaps Santa Fe, New Mexico.[39]

The morning of March 27, 1894, Douglass traveled north on the stage from Phoenix to Prescott, recording his observations on the local weather, especially clouds. For one of the few times on his trip, he had no room to carry the telescope and accessories with him. Therefore he shipped his instruments from Phoenix to Prescott by rail express. While Douglass himself traveled from Phoenix to Prescott, the telescope went by rail from Phoenix back to Tucson, and then to El Paso. From there, it journeyed up the Rio Grande valley to Albuquerque, west to Ash Fork and south to Prescott, a total distance of some 1100 miles. Remarkably, Douglass arrived less than twenty-four hours before his equipment.[40]

On the day Douglass reached Prescott, Lowell telegraphed that two observations would be sufficient for both Prescott and Flagstaff. Once again, the weather foiled carefully made plans. For two days, clouds around Prescott promised rain, making telescopic observations worthless. Douglass examined various possible sites in the area, concluding that three low hills south of town would be the only likely location for observations. When the weather finally cleared on March 31, he found only fair seeing. After two more nights of the same quality, Douglass departed for Flagstaff on April 3.[41]

When his train rolled into the forest-shrouded town of Flagstaff that night, Douglass noted that while the stars clearly twinkled, no clouds interfered. Beginning on the evening of April 4, he took several complete series of observations at various Flagstaff sites. On a low mesa west of town, Douglass made ten sets of ten sightings each during a ninety-minute period, obtaining average values ranging from 4.6 to 8.

On April 6, Lowell telegraphed Douglass to check the summit near Spring Station, six miles northwest of town. Douglass examined this and various other sites and noted that, by and large, the seeing in Flagstaff remained slightly superior to that in Prescott. On April 14, he recorded fifty observations which convinced him (and Lowell) that the observatory could be located in this northern Arizona lumbering town.[42]

After receiving Douglass's reports of the previous two nights' observations, Percival Lowell telegraphed his assistant on April 16 that Flagstaff would be the site of the new observatory. In a letter, he said that site eleven, a low hill west of town, looked the best so far, but Douglass was to use his own judgment in selecting locations. Five days later, however, Lowell instructed Douglass to begin work at site eleven, even though his assistant suggested that air currents would be less of a problem on one of the neighboring hills. The approaching Martian opposition led Lowell to disregard Douglass's advice in the name of expediency. An adequate site immediately available was of greater value than a potentially excellent site not yet located.

Although based on meager evidence, Lowell's decision seemed the only guarantee of early Martian observations. Douglass thus began preliminary work on site eleven, less than a mile west of town, breaking ground on 23 April.[43] His apprenticeship clearly ended, the young astronomer was about to embark on the first major step of his scientific odyssey, a step which would involve him in one of the most exciting debates in late Victorian astronomy.

2

The Canals of Mars

FOR THE NEXT EIGHT MONTHS, the Lowell Observatory in Flagstaff completed its physical plant, made the first use of its astronomical equipment, and began the intensive study of Mars for which it had been founded. In all these endeavors, Douglass contributed significantly. Readying the site while Lowell and Pickering remained in Boston, he supervised the construction of the housing for the telescope and made the important early contacts with the Flagstaff community. During the months of observations, Douglass remained the staff's principal investigator, spending far more time in the observer's chair than either Lowell or Pickering. The observational skills of A.E. Douglass and the clear atmosphere of the American Southwest combined to add significant knowledge to the study of astronomy.

Douglass's early career paralleled the beginning of a new era in observational astronomy. The traditional astronomical technique of peering through a telescope and sketching one's view, which had been the sole method since the early seventeenth century, was giving way to more sophisticated practices. Students of planetary phenomena were slowly abandoning artistic drawings in favor of photographic plates which, although still rather primitive, eliminated the more obvious shortcomings of drawings made from memory. During the second half of the nineteenth century, stellar and nebular astronomy also advanced, with significant use of the spectroscope to identify the chemical elements contained in these phenomena and the motions their spectra frequently indicated. The new astronomy required sophisticated equipment and a suitable climate in which to employ it. Douglass's trip to locate a site for Lowell's observatory represented a critical awareness of the atmosphere as an integral part of any astronomical investigation.

Although a site had been selected, establishing the observatory required many other endeavors. While Douglass had surveyed sites in Arizona, W.H. Pickering designed and constructed in Cambridgeport, Massachusetts, a dome to protect the telescope and other sensitive instruments. Built of wood and canvas, the dome was broken down into sections and shipped to Tucson, arriving on March 21, 1894.[1] Pickering and Lowell also acquired the necessary instruments for the observatory. Pickering obtained the loan of an eighteen-inch refracting telescope, the largest instrument available at the time, recently completed by the experienced Pittsburgh lensmaker, John A. Brashear. From Harvard he borrowed a smaller telescope, of twelve-inch aperture, while Lowell leased the instrument's clockwork and other accessories. Boston instrument-makers Alvan G. Clark and Sons modified the equipment to fit their custom-made dual mounting for the two telescopes. This special mounting saved space, expense, and time, since the Clarks could construct the dual mounting with less material and labor than required for two separate devices. Other accessories included a spectroscope loaned by Brashear, photographic apparatus supplied by Clark, and a polariscope and other equipment purchased by Lowell.[2]

On April 23, Douglass began breaking ground at site eleven, on what soon became known as Mars Hill. Using local material and labor, Douglass erected the framework for the observatory base out of twenty-four cedar posts, braced and set deep in the ground. Once this foundation had been completed, Douglass moved to preparations for the instruments. On Monday, May 7, workers placed the pier, a heavy iron tube to hold the telescope mounting, in a deep hole which had been partially filled with broken stone and portland cement. Early the next week, workmen finished lowering the iron tube into place, using two forty-foot pine spars as levers. To complete the finer adjustment, two men stood in the hole and moved the tube while Douglass watched through a theodolite to make sure of the pier's accurate position. Following final adjustment, plaster of paris was poured inside the tube and allowed to harden. Workmen then added mortar, rock, and sand around the foot of the pier and threw in heavier material to fill the hole.[3]

Douglass next supervised the preparation of the base for the dome. He first attached the walls to the cedar posts, using thin boards to keep the outside and inside temperatures approximately equal. This precaution was necessary to maintain the telescopes in proper adjustment by minimizing the expansion and contraction of their lenses due to temperature fluctuations. Douglass then constructed a "live-ring," a ring of eighteen beveled wheels running between hardwood tracks on top of the base wall. The dome, which had been freighted from Tucson by rail, would be placed on top of this device. The tracks caused little difficulty, but the complete assembly failed to operate smoothly, making movement of the dome almost impossible. Douglass removed and realigned half of the wheels, but the problem remained. He finally had to disassemble the ring completely and remove and reset every wheel, delaying completion of the observatory building.[4]

The city and people of Flagstaff generously assisted in the completion of the observatory. A week after groundbreaking, the city started building a road to the site and formally agreed to give outright title to ten or fifteen acres of land for the installation, the exact amount to be determined by Douglass at a future date. The Atlantic & Pacific Railroad delivered free of charge the telegraph poles supplied by Western Union at cost. Despite this assistance, the outlay alarmed Lowell. "The building expenses of the dome etc. in Flagstaff threaten to equal the magnifying power of [the] 18 inch," he wrote Douglass. "Keep an eye on them, that they distance not the ephemeris."

The observatory building and dome were completed when Pickering arrived on May 20 with the eighteen-inch Brashear lens, ready for mounting in the telescope. Lowell reached Flagstaff on May 28, anxious to use his newly completed facility, but the weather refused to cooperate. Cloudy skies prevented observations with the Harvard twelve-inch until May 31. The first use of the larger Brashear instrument did not take place until the next night.[6]

The astronomers at Lowell Observatory concentrated on Mars from late May until early April of 1895. Lowell spent long hours at the telescopes in June, in late August, and in October and November, with Douglass and Pickering handling the remainder of the work at the observatory. From June until early December, when Pickering and Lowell departed, the astronomers spent almost every night in the dome, but bad weather severely reduced observations during the winter. During the initial observing season, the staff completed a wide variety of observations, including 917 drawings of the planet's surface, fifty-seven measures of the polar cap, 464 micrometric measurements of the polar and equatorial diameters, and 736 determinations of irregularities on the terminator, the line separating day and night on the Martian surface.[7]

The Search for Life on Mars

The Flagstaff observations provided support for Lowell's theories concerning Mars. The *canali* discovered by Schiaparelli, and mistranslated into English as "canals," were the starting point for Lowell's ideas about the red planet. To support the theory that these lines represented watercourses, it was necessary to prove the existence of moisture and an atmosphere on Mars. Observations at the Lowell Observatory of the south polar cap showed that the cap seemed to shrink in size as the Martian seasons progressed from spring to summer. At the same time, the dark surface markings, including the canals, seemed to darken. To Lowell, this paralleled the same phenomenon as observed on Earth and suggested that vegetation existed on Mars. As the polar caps melted, water flowed through a system of canals to irrigate the vegetated areas of the planet.

Mars did not, however, present a lush tropical paradise. Most of its surface remained "desert," as indicated by the planet's reddish color. The scant available water seemed to be concentrated at the poles, and would have to be brought from them to sustain life. Intelligent beings on Mars would be concerned primarily with irrigation and their necessarily large-scale systems would be visible from Earth. In Lowell's words, the observed lines, spots and triangles on the Martian surface "are just the features a gigantic system of irrigation would present." The conclusion (to Lowell, at least) was inescapable: Mars presented a generally habitable surface, inhabited "by some form of local intelligence."

Douglass's work during the 1894–95 opposition proved crucial to the development and support of Lowell's theories. His raw observations and interpretation of the collected data supported his employer's formulation and defense of the Mars theory. A very important part of this work was the determination of the planet's shape. Between July 6 and November 21, 1894, Douglass made 403 observations of the Martian disk, measuring the polar and equatorial diameters with a micrometer attached to the telescopes. These measurements, when reduced, gave a polar diameter of 9.″32 and an equatorial value of 9.″37. From these values, the polar compression of the planet was calculated to be 1/190, indicating a shape roughly similiar to that of the earth. Douglass also discovered the existence of a twilight arc, a phenomenon caused by the refraction of sunlight through a planet's atmosphere. This observation further supported the generally accepted existence of a Martian atmosphere, however tenuous that gas envelope might be.

Douglass's work reinforced the view that the necessary preconditions for life existed on Mars. From its size and its limited atmosphere, many astronomers had concluded that Mars had formed before the earth. If Mars had developed similarly to Earth, a logical conclusion could be drawn that life had appeared on the red planet before it had appeared on Earth. Assuming this to be the case, any intelligent life on the planet probbably had a longer period of evolution and, consequently, a higher civilization than its terrestrial neighbors. If this had been the path of Martian development, the planet's inhabitants probably possessed a technology enabling them to construct the intricate canal network on the planet's surface. The similarities discovered by Douglass between the oblateness of Mars and that of the earth reinforced this idea of a more rapid evolutionary history for the red planet. The polar flattening of Mars could be explained by positing a prior molten state for the planet, equivalent to that of Earth, and a more rapid cooling because of Mars's smaller size; thus, Mars presented a model of Earth's future development, and suggested that Martians might exhibit a more advanced civilization than the inhabitants of their sunward neighbor.[9]

The investigations of the Martian canal network consumed much of Douglass's time and energy. The lines on the Martian surface, which early impressed Douglass "with their great length and straightness, and geometrical arrangement," remained the most dramatic and intriguing aspect of the planet's features. During the first few months of observations, Douglass remarked that canals could also be seen in the dark areas of the planet, apparently indicating that watercourses covered the entire Martian surface, including the polar zone. While in Arequipa in 1892, Douglass and Pickering had observed streaks in the dark areas of Mars, a discovery repeated by J.M. Schaeberle at the Lick Observatory the same year. None of these astronomers, however, had believed the streaks to be canals.

At Flagstaff, Douglass discovered that the canals in the dark regions connected with those of the "desert" areas, showing a complete canal system extending from the poles across the light areas and continuing to the dark regions. The fifty-six canals discovered by Douglass in the dark areas appeared in every way similar to those in the light regions, providing Lowell with valuable support for his theory of irrigated Martian vegetation and the civilization which produced it.[10]

Douglass also investigated the rare phenomenon of doubling (or gemination, in which two parallel canals appeared in a position previously occupied by only one), first observed by Schiaparelli. Although seldom visible, this process attracted wide attention, but no cogent explanation. Douglass surmised that the many parallel canals in the dark regions indicated that the supposed gemination was only the filling up of adjacent canals. He thought this "not an unlikely hypothesis," but lack of observations made confirmation impossible, and Douglass remained skeptical of the phenomenon. Only with exceptionally good seeing and often only at twilight was gemination visible, and then the phenomenon appeared rarely.[11]

The canal research of Douglass bolstered Lowell's Martian theories. The discovery of canals in the dark regions, the recognition that canals darkened rather than broadened as the Martian seasons progressed from spring to summer, and the confirmed interconnection and geometrical nature of the canals all enabled

Lowell to put forward a significantly new view of the Martian surface. Proceeding from the great scarcity of water on Mars, and the need of any intelligent civilization for an irrigation network, Lowell concluded that the Martians had constructed just such a system. He also believed that the dark lines on the Martian surface were artificially irrigated country surrounding the waterways, and the larger dark areas represented vegetation watered by the canals. Accepting the lines as vegetation eliminated the necessity of explaining how an arid world could support canals fifteen to thirty miles wide, the minimum width necessary to enable terrestrial observers to view such examples of Martian engineering.[12]

Douglass's observations of the Martian terminator provided further support for Lowell's theories. During the 1890 and 1892 oppositions, several astronomers had noted irregularities on this line, either light spots in the darkened part of the planet or dark spots in the portion of Mars in sunlight. During the 1894–95 opposition, Douglass and the staff at the Lowell Observatory observed more than 700 such irregularities. Many astronomers explained the phenomena in terms of surface features such as mountains or craters and valleys. As on the moon, a bright spot in the dark area (called a projection) could be caused by the sun's rays hitting a mountain peak whose base remained in shadow. A dark area in the portion of Mars illuminated by the sun could be caused by a shadow of a mountain or crater rim.[13]

Douglass argued that terminator irregularities were not caused by mountains and valleys. A mountain peak could only be illuminated for a certain time until the planet's rotation would carry it out of the sun's rays. Douglass also suggested that the vast changes in position of such phenomena indicated that an atmospheric explanation for the irregularities would be more convincing. The long continuity on some nights, the great reflecting power of projections, and their irregular movements all pointed toward large cumulus-like clouds as the cause of such phenomena.

Douglass's explanation of projections developed in part from his observations in November of 1894. On the night of the twenty-fifth, he saw a large projection appear, rapidly lengthen parallel to the terminator, and suddenly disappear a half hour later. The projection reappeared the next evening, but nearly nine degrees farther north. For the rest of the second night, the projection dissipated and reformed at irregular intervals. It could not be located on the twenty-seventh.

Douglass speculated that the phenomenon was a cloud. The fact that the projection suddenly disappeared on the first night, without any lessening of its size, cast doubt on the mountain explanation. When the projection reappeared the next night well north of its previous position, Douglass was sure that the projection represented an atmospheric phenomenon. He calculated that the cloud varied in width from twenty-four to fifty-eight miles, extended approximately 142 miles in length, and floated at a minimum height of twenty miles above the surface, thought possible because of the thinner Martian atmosphere. The presence of clouds and moving air masses further suggested a Martian atmosphere, which was necessary for the existence of life forms. Depressions, the dark irregularities on the illuminated side of the terminator, Douglass suggested, represented surface features which caused a decrease in reflecting ability. Douglass's terminator work led Lowell to construct a Martian climate which, because of its aridity, would explain the need for an intricate canal system, but would also permit the existence of the civilization needed to construct that system.[14]

Douglass's activities were not totally concerned with making observations. As a member of the Lowell Observatory staff, whose director had quickly become identified as the most outspoken proponent of Martian civilization, Douglass was drawn into the debate surrounding the possibility of life on the red planet. Even before the 1894–95 opposition, a number of contrasting ideas had been published in the scientific press. The canals, particularly, came in for vituperative debates. Henry H. Bates, a member of the Astronomical Society of the Pacific, dismissed the idea of waterways on Mars in 1894. He believed that the planet had no surface water because it was far too cold. The polar caps were frozen carbon dioxide which would wax and wane as had been observed, and might possibly result in a liquid; but since carbon dioxide usually sublimed from a solid to a gas, this remained only a slight possibility. In short, Mars "is chilled to the center and is cracked through and through in many directions, as the immense cañons, absurdly called canals, evince."[15]

Edward S. Holden, director of the Lick Observatory, thought Lowell's ideas about canals were equally presumptuous.

How could engineers build "canals" no less than fifty miles wide, which appeared and disappeared, and which often stretched 1000 miles? "[S]hould we think these waterways were built by intelligent beings like ourselves, or by madmen?" he asked. One of his colleagues at Lick, W.W. Campbell, questioned the existence of a Martian atmosphere, citing his own spectroscopic studies which indicated that the spectra of Mars closely paralleled those of the moon, with no extensive atmosphere. Even before Lowell's first publications from his research, therefore, serious astronomers had developed plausible alternative hypotheses.[16]

Although Lowell, Pickering, and Douglass focused on Mars, Douglass also swung his telescope to other worlds. With Pickering, he studied Jupiter, the largest member of the solar system, and its four principal satellites. They estimated the depth of the Jovian atmosphere at between 2800 and 3750 miles and calculated the polar oblateness at 1/16.11, indicating a significant bulge at the planet's equator, and agreeing with later estimates.[17]

Douglass was particularly interested in determining the characteristics of the four largest Jovian satellites: Io, Europa, Ganymede, and Callisto. The shape of Io proved interesting, appearing to be egg-shaped and revolving end over end. Europa also displayed polar flattening, but to a less dramatic degree. A north polar cap and other surface markings appeared on Ganymede, but the most astounding discovery was that the satellite had three unequal axes, a most unusual occurrence. Callisto, the most distant of the moons, also presented surface markings and the prolate spheroid shape, but was thought to keep the same face toward Jupiter at all times. The observed markings were confirmed by NASA's Voyager missions in 1979, but no major irregularities were found in the satellites' shapes.[18]

During these months Douglass made use of his rare free time to design a new astronomical instrument, a large dome-telescope. He proposed a telescope 100 feet long mounted in a hollow sphere which would float in water for support. In January of 1895, he sent Lowell a draft of an article on this invention and asked for his employer's "frank criticism as to whether I shall be made ridiculous by publishing such a thing." In a second letter Douglass asked if Lowell had formed opinions on the matter and suggested that the advice of Warner and Swasey, the Cleveland firm which had built the Lick Observatory, would be valuable. Lowell answered that their opinion would be worthwhile, but added that floating a telescope on water had been tried unsuccessfully at Nice, France, and at Lick.

Douglass published his article in the May 1895 issue of *The Astrophysical Journal*. His final design called for a seventy-two-inch telescope mounted in a sphere of thin steel 100 feet in diameter. This "dome" floated on water in a well of adequate depth. Designed in an attempt to improve the steadiness of movement and rigidity of support, the proposed observatory would have axis bearings so that the sphere could float higher or lower, depending on the weight of the observer and his accessory equipment. Douglass also sought a system for automatic balancing. When the sphere was thrown out of balance, the strain on the driving-gear wheels could cause electrical circuits to add water to the tank to rebalance the structure. After publishing his study, Douglass abandoned further work on the idea.[19]

Although few residents possessed clear ideas about the research on Mars Hill, Flagstaff displayed great interest in the Lowell Observatory. As the building neared completion, Douglass had published a letter of appreciation in the local newspaper, the *Coconino Sun*. He thanked Flagstaff for its help and announced that after the Mars opposition, the observatory would quite likely be opened to the public at various times, providing an opportunity to look through the telescopes. Beginning in August of 1894, the staff began scheduling open evenings, which were always well attended. Although Mars and the moon remained the most popular objects, Douglass showed other planets, as well as stellar and nebular phenomena. In return, the observatory staff was a popular addition to the social activities of the town. The Flagstaff press reported various astronomical events at or concerning the observatory, and frequently gave front-page space to reprinted articles about Mars.[20]

The determining factor in locating the Lowell Observatory at Flagstaff had been the atmospheric properties of the northern Arizona plateau. Throughout the summer and fall of 1894, the weather behaved admirably for astronomical observations, but after Lowell and Pickering returned east in December, winter weather set in, with snow, rain and cloudy skies. Douglass wrote Lowell on December 28: "Last night as usual was cloudy. Tonight will be likewise." "Flagstaff in the winter is a fraud, apparently," Lowell replied on January 5, 1895. "You have my condolences."[21]

Despite the cold weather, Douglass regularly climbed Mars Hill from his boarding house to inspect the observatory. In early January, while closing the shutter in fear of an approaching snow storm, Douglass was "rewarded" with a slight clearing of the weather, and made a few observations in seeing no better than three. On one occasion, when the weather cleared, he trudged through the snow for a night's observation. Recent snowfall had been so heavy that Douglass could enter the dome only by climbing a ladder braced against the outside of the building and easing through the shutter. Once inside, he could not turn the dome. The snow conditions, however, did have a constructive result, for Douglass learned how to use his snowshoes as a toboggan to slide down the hill. "In fact," wrote Douglass in explaining his technique to Lowell, "I may say that my chief astronomical work is studying the 'ski'—quite as it should be."[22]

As 1894 drew to a close, Douglass could reflect on eight months of significant achievement. The Lowell Observatory had been completed in Flagstaff and had immediately established itself as the center of the continuing investigation of Mars. As principal assistant and chief observer at the facility, Douglass contributed to the development of Lowell's imaginative theory concerning life on the red planet and proved himself a valuable addition to the staff. The end of the year, however, brought a disheartening change of fortune as the Flagstaff winter ran its course. Douglass could do nothing but gaze at the cloudy sky and shake his head in frustration and disappointment. He spoke for the entire observatory when he wrote, "Another cloudy, rainy, snowy night has just passed and a repetition is in prospect for tonight. When will our weather change?"[23]

The next few months in Flagstaff presented no improvement and precluded regular work at the Lowell Observatory. In March of 1895, Douglass received a disturbing letter from Lowell, precipitated by the former's frustrating experiences with the northern Arizona climate. "The seeing seems to be so perpetually poor now," wrote Lowell from Boston on the fifth, "that I see little use in keeping up the observatory longer." He added that he planned to find another location for his facility, possibly near Mexico City, and instructed Douglass to return the Brashear and Harvard instruments to their owners by the first of May. The dome was to be dismantled and shipped to Boston and Douglass was to sell the ordinary furniture and other equipment.[24]

The Quest for a New Observatory Site

By mid-April, Douglass had shipped the most delicate parts of the instruments east, and had officially closed the Lowell Observatory. He also attempted to smooth over hurt feelings in Flagstaff. The town took great pride in the Lowell facility (frequently advertising it as a tourist attraction) and was disappointed with the closing of the observatory. Douglass informed residents that within six years the most important work on Mars would be accomplished, and that if the Lowell Observatory returned to the United States, it probably would be placed at Flagstaff. He also convinced Lowell to leave the dome on its foundation rather than ship it to Boston.[25] This procedure would save money, convince Flagstaff of the observatory's good intentions, and provide the astronomers with a reserve site if the proposed Mexican location proved inadequate.

Douglass concurred with Lowell's suggestion that the Mexican tablelands might prove a good site, but requested a leave before joining the new project. He informed his superior that he had to visit Edward C. Pickering, Director of the Harvard College Observatory, to reduce at least part of the earlier Arequipa observations. Two or three months at Harvard, he insisted, were imperative, after which he would be free to "start off once more." Douglass also mentioned financial difficulties. "Living quietly in a city like Cambridge seems to be less expensive than living in foreign parts," Douglass stated, "and my salary of $800 [per annum] has not been sufficient for my personal expenses during the past year." He told Lowell that a raise was not a condition of remaining, as he knew the advantages of a professional connection with the Lowell Observatory, but the financial problem remained a consideration. The determining factor, he assured Lowell, was the need to get some of the Arequipa work ready for publication. Douglass would then renew his connection with Lowell "at once."[26]

While waiting for word on his request, Douglass began drafting plans for a comprehensive survey for the new site of the Lowell Observatory. He proposed to start in Flagstaff in mid-winter and work his way south to the equator. Observations would be made every five to ten degrees of latitude at similar altitudes in the mountain ranges. On the Mexican plateau, he hoped to take observations at various altitudes to test the effect of atmospheric currents. Douglass also suggested that an observatory on an island with a mountain high enough to reach above the trade winds might be considered. Anticipating the erection of such a facility by seven decades, he noted the Sandwich (Hawaiian) Islands as a possibility.[27]

Lowell's response to Douglass's request came on April 22, bringing W.H. Pickering's shocking estimate that Douglass's work on the Arequipa material would require eight months. With this news Douglass offered to hire people at his own expense to do the mathematical calculations, and also asked Pickering to intercede in his behalf with Lowell. The principals arranged a solution. Douglass could accept Lowell's proposition, providing that Harvard could borrow his services in the near future. Douglass made a final check of the observatory building in Flagstaff and prepared to leave for Mexico with Lowell's six-inch refractor. A pleasant surprise came on May 4, when he received a telegram from Lowell saying that his salary had been increased to $1500.[28]

Leaving Flagstaff at dawn on May 9, Douglass spent the next month surveying possible observatory sites between Mexico City and Chihuahua. The sites proved disappointing, with seeing seldom better than five. Even the Mexican National Observatory at Tacubaya, southwest of Mexico City, presented marginal atmospheric conditions. Douglass returned to Boston on June 25, resuming work at the Harvard Observatory to prepare the Arequipa data for later publication. Over the next four months, the massive collection of research material slowly took on a more orderly appearance, but final preparations for publication remained far in the future. In early November, with significant work accomplished at Harvard, Douglass again left for Mexico. His second site survey brought little encouraging new information and he returned to Boston in mid-December. The next Martian opposition was less than a year away, and still the Lowell Observatory had no suitable location.[29]

Douglass returned to his labors at Harvard in January of 1896, further organizing the results of the Arequipa expedition. At the same time, Lowell traveled to the Sahara Desert to check on possible observatory sites. By early March, he had abandoned this project, finding the great desert "mediocre" and cabling Douglass to return to Mexico as soon as he finished his Arequipa work. Douglass remained very busy at Harvard throughout the summer of 1896, trying to complete his Arequipa responsibilities. Warning that "Time gallops," Lowell asked his assistant in June to estimate the end of his Arequipa endeavors, but by July he could wait no longer. He ordered Douglass to leave for Flagstaff by July 5.[30]

Dutifully, Douglass left for Flagstaff, arriving there on the eleventh. The following week, men began unloading the machinery and apparatus for Lowell's new telescope, to be placed in the old dome on Mars Hill for tests and adjustments. Purchased at a cost of $20,000, the twenty-four-inch instrument was one of the largest refractors in the United States. On July 22, Lowell arrived with the lens and its maker, Alvan G. Clark. The astronomers installed the lens quickly, beginning observations the next night and continuing through the summer from the old hemispherical dome. Lowell and Douglass were joined by three new astronomers from the University of Chicago: David A. Drew and Wilbur A. Cogshall, two graduate students, and Dr. Thomas J.J. See.[31]

In early November, Douglass traveled again to Mexico. Taking advantage of the offer of assistance from Felipe Valle, Director of the National Observatory, Douglass ultimately chose the Tacubaya area because of its accessibility by both rail and wagon. The chosen site lay one-half mile south of the National Observatory on land whose owners allowed the free use of their

Lowell Observatory, Flagstaff, Arizona, ca. 1901. The unusual cylindrical dome was designed for the Lowell station at Tacubaya, Mexico, and was shipped to Flagstaff after the 1896–97 Martian opposition.

property. Once again, the Lowell Observatory found itself forced to choose a marginal location.[32]

While Douglass saw to details in Mexico, Lowell ordered a larger dome constructed for the new station, contracting with Godfrey Sykes, in Flagstaff, to do the work. By November, Sykes had designed and built a forty-two-foot cylindrical dome made entirely of wood and weighing more than four tons. On November 9, workmen carefully removed the lens from the telescope and packed it for shipment. Lowell carried the glass with him when he left for Boston the next day. The observatory dome and equipment were then loaded on a railroad car, which left Flagstaff November 22.[33]

In Mexico, Douglass continued his many activities. He obtained reduced rates for shipping the observatory equipment on the Mexican Central Railroad and started construction of a base at the Tacubaya site, completing it by December 5.[34] Three days later, fifteen Mexican laborers under the direction of Godfrey Sykes, who had accompanied the boxcar, began moving the equipment from the car to the site. By noon, the entire seventeen tons of material had been unloaded without mishap. Mounting the telescope and dome at Tacubaya took two more weeks, despite Lowell's telegram to "hurry things please." Great delay was caused by difficulties in adjusting the double track upon which the dome would revolve. Once the adjustments were completed, the roof pieces were put in place, the canvas shutters installed, and interior necessities such as the floor and observing chair secured. Lowell arrived with the telescope lens and installed it on December 28.[35]

Observations of Mars began on December 30, well after the opposition (which decreased the number of useful sightings), and continued through March 26, 1897. Conclusions of the 1894–95 opposition in Flagstaff were "confirmed." The staff, after studying again the north polar cap, concluded that a noncontinuous cloud layer, in addition to snow and ice, formed the cap. The Martian canals again attracted the most attention, with the staff raising the number observed to 214. The astronomers calculated the area covered by canals at almost ten million square miles, slightly less than one-fifth of the total surface area of the red planet. The debate over the "double canals" continued, but Douglass (and, to a lesser extent, Lowell) questioned their reality, chiefly because of their infrequent appearance in clear atmosphere and their greater visibility during periods of poor seeing.[36]

Douglass also studied the major satellites of Jupiter, confirming the surface markings on all four. The ellipticity of Io appeared to have increased noticeably since he studied it at Arequipa, leading him to believe that there had been a decrease in its diameter. As the diameter decreased, the rotational velocity would increase, due to the conservation of angular momentum. Increased rotational velocity would then increase polar flattening. Less severe changes appeared on Europa, while Ganymede and Callisto appeared to present the same face toward Jupiter at all times. As Ganymede had surface markings like the "canals" of Mars, Douglass suggested a similar explanation, that they were permanent surface features, which caused W.H. Pickering to question the young astronomer's wisdom in this matter. How could he reconcile the permanent canals on Ganymede with the changes in the satellite's shape they had seen in Arequipa? Pickering felt it "rather difficult to conceive of belts on such a body, but I think permanent canals increases [*sic*] the difficulty materially."[37] Douglass returned to his speculations.

While in Mexico, Douglass also pursued one of his favorite diversions, mountain climbing. Nearly thirty miles southeast of the Lowell Observatory rose Popocatépetl, the country's most famous mountain. A challenge to any climber, the peak's summit soared almost 18,000 feet. As soon as spring weather permitted, Douglass made plans for his ascent, which included Cogshall and, for part of the way, T.J.J. See and his brother. On April 13, 1897, they spent approximately six and one-half hours climbing from base to summit. At the higher elevation, frequent rests became necessary. The climbers found that one-minute rests every twenty or thirty steps replenished enough of their energies to continue. Douglass carried a barometer throughout the climb, and on the summit made altitude observations with his instrument. The descent of Popocatépetl took less than two hours.

A week after the ascent, Douglass received a congratulatory telegram from Lowell in Boston: "Bravo upon Popocatépetl! First as usual." Douglass soon planned another expedition, this time to 18,700-foot Pico de Orizaba, some 180 miles east of Popocatépetl. Again taking Cogshall, he began the ascent on April 28, but failed to complete the climb to the summit, fatigue and altitude overcoming him at 17,300 feet. Cogshall, however, completed the ascent the next day.[38]

Douglass's mountain-climbing expeditions in many respects proved more successful than the astronomical work of the transplanted Lowell Observatory. The end of the 1896–97 Martian opposition marked the end of Lowell's connection with astronomy in Mexico. Atmospheric quality and accessibility proved inadequate to justify a continuation of the Lowell Observatory in Mexico. Despite the potential difficulties of Flagstaff winters, Lowell's previous northern Arizona location seemed quite attractive. The observatory staff refocused their attention to the north, anticipating another move.

3

Conflict and Isolation

ON APRIL 22, the *Flagstaff Sun-Democrat* announced under a headline of "Skylight City Beats Mexico" that Percival Lowell had decided to bring his observatory back to Flagstaff. In fact, the telescope was already on its way to the northern Arizona town. Lowell, however, would not be returning with his instrument, having suffered an attack of nervous exhaustion soon after returning to Boston from Mexico. This illness incapacitated him for four years. William L. Putnam, Lowell's brother-in-law, assumed direction of the observatory's business affairs in Boston, designating Douglass to supervise the Flagstaff installation.[1]

The reestablishment of the Flagstaff observatory proceeded smoothly under the able direction of Godfrey Sykes. Most of the equipment had been reassembled by May 6, with the dome used in Mexico erected on Mars Hill to enclose the twenty-four-inch refractor. Douglass, Cogshall, and See arrived two days later. By mid-day, the lens had arrived and observations began. Then suddenly Douglass himself fell ill with what a Flagstaff newspaper called "quite a severe attack of rheumatism." To recover from this ailment, he went to southern California for an extended visit, leaving See in charge of the Lowell Observatory.[2]

While recuperating from his illness, Douglass stayed with an aunt in San Diego. During the summer and early fall, he accomplished some preliminary work on the Mexican observations and visited the Lick Observatory southeast of San Francisco. In November, See indicated a desire to travel east for the winter, which would have left no experienced astronomer at the observatory. Putnam telegraphed Douglass to return to Flagstaff as soon as possible to take See's place, and he obligingly returned on November 17, 1894.[3]

See's departure was a relief to the staff of the Lowell Observatory. Despite his impressive credentials (Ph D., University of Berlin, 1892) and three years' experience at the University of Chicago under George Ellery Hale, See proved to be a very difficult colleague. His abrasive personality, arrogance, and penchant for using others' work as his own prompted the resignations of Drew and Cogshall, the latter replaced by Samuel Boothroyd of the University of Chicago. The assistants wrote frequently to Putnam, detailing the personal insults to which See subjected them. Lowell and Putnam were disturbed by the situation, writing Douglass on June 14, 1898, that if it came to a choice between See and Douglass, the former would go. Putnam said he would talk to See when he came to Boston. It would be "just as well not to talk to many people about Dr. S.," he confided.[4]

In late June of 1898, Douglass sent Putnam an uncharacteristically harsh judgement of See. "Personally, I have never had such aversion to a man or beast or reptile or anything disgusting as I have had to him," he wrote on the twenty-eighth. "The moment he leaves town will be one of vast and intense relief and I never want to see him again. If he comes back, I will have him kicked out of town." Within a week after the arrival of Douglass's letter in Boston, Putnam dismissed See. Convinced that See's personality and questionable integrity made him unfit to hold positions of responsibility, Douglass attempted to block his appointment to the staff of the U.S. Naval Observatory. Despite Douglass's opposition, the observatory hired See as a professor of mathematics, a post he held for many years.[5]

The work at the observatory continued. Douglass made observations of Venus, Mercury, and the Jovian satellites during the summer of 1897. He also studied Mars during the 1898–99 opposition, but could discern only 132 canals. He examined Io, the innermost satellite of Jupiter, confirming its period of rotation at twelve and one-half hours (quite removed from current estimates of 42.5 hours) and employing a comparison scale of ellipticities to determine its polar compression.[6]

Douglass also made improvements to the observatory while serving as acting director. He had the dome covered with wooden sheathing and a layer of thin steel. He later installed a floating dome during the spring and summer of 1899, one of only two in the world. On top of the observatory walls, workmen constructed a circular trough twenty inches wide, twenty inches deep, and forty feet in diameter. Lined with lead, it held a salt-water solution to increase buoyancy and prevent freezing in winter. Twelve pontoons, each ten feet long and covered with tin, were fastened on the underside of the dome itself, allowing the entire structure to float and making it easier to move.[7]

The search for better observatory sites continued as well. During the first summer after the return to Flagstaff, Douglass recommended Hawaii, but Lowell advised against the islands because of their isolation. South America was also suggested, but Lowell's continued illness prevented serious attention to the matter. Nonetheless, Douglass continued to collect data on different localities, none of which impressed him greatly. On a trip east in the fall of 1898, he stopped briefly in Santa Fe and Denver, and made cursory tests. The Santa Fe *New Mexican* took special notice of Douglass's visit, concluding that the Lowell Observatory would be moved there. The Flagstaff *Sun* reported this story with some alarm, but a week later assured its readers that no move was contemplated.[8]

Despite Lowell's illness, the observatory resumed its former operation in many aspects. Local newspapers announced evenings of public viewing and provided background information for potential visitors. The press also reported on prominent figures who visited the observatory. During June of 1898 E.E. Barnard, the discoverer of Jupiter's fifth moon and a veteran of the staffs of both the Lick and Yerkes observatories, inspected the Lowell station. On June 4, Douglass took Barnard on a leisurely ascent to the summit of 12,340-foot Agassiz Peak, giving the visitor an unparalleled view of the observatory below.[9]

A.E Douglass (at right) and George A. Waterbury at the Lowell Observatory, Flagstaff, 1898. Waterbury was a frequent companion on Douglass's many mountain-climbing excursions in the Flagstaff area.

Lowell's illness prevented any significant changes in Flagstaff. The most important activity, Putnam wrote, was to keep the observatory in efficient condition. This required Douglass, as senior assistant, to remain in Flagstaff as much as possible and oversee the disposition of the observatory's $6000 annual budget. During the years of Lowell's illness, Douglass directed little more than a caretaker staff in Flagstaff, although in mid-1899 Lowell asked him to "search *carefully*" for the ninth satellite of Saturn, discovered the previous year by Harvard's W.H. Pickering. He had no success.[10]

Douglass spent long hours processing data gathered by the observatory. He assisted with the observatory publications, and published his own contributions. Two volumes of the observatory's *Annals* were published in Boston during Lowell's absence, the first in 1898 by Houghton, Mifflin and Company, and the second in 1900 by Harvard University. After Lowell's collapse, Douglass assumed the task of editing the first volume in October of 1897. Most of the data in the volume had already been arranged and interpreted, so he corrected page proofs, prepared the index, and saw the work to the printer. On the second volume, Douglass did almost all of the work, reducing the raw data, arranging them, and writing the analyses. Although Lowell urged him to complete the task within four months (and expressed grave disappointment when this was not accomplished) Douglass labored for almost an entire year in both Boston and Flagstaff on the volume which appeared in late 1900.[11]

In addition to his endeavors with the *Annals,* Douglass wrote several articles on the work of the observatory, often at the request of Putnam. In 1898, when E.E. Barnard, program director for the astronomical section of the American Association for the Advancement of Science, requested an article on the Lowell Observatory, the letter was forwarded to Douglass, who prepared a paper for the association's September 1898 meeting. He contributed frequently to the widely read *Popular Astronomy*, providing the journal with such impressive material that the editor forwarded two subscriptions to the observatory at no charge. Douglass's contributions were valued both for their interesting accounts of planetary research (of less interest to most astronomers than stellar and nebular work, but more popular with amateur observers) and for their semipopular style. The editor encouraged Douglass to send more: "We esteem the matter from Lowell Observatory among the freshest and best that we get from any source."[12]

Many of Douglass's publications described the results of his research. In *Popular Astronomy* he discussed a specially modified globe used to record planetary markings and, in the same issue, provided a detailed discussion of the interrelationship between "Atmosphere, Telescope, and Observer." Focusing on atmospheric currents that hindered good seeing, Douglass ascribed their cause to a combination of many factors, including the settling of cold air at night, nearby mountains or hills, snow, cloud condensations, and streams of air at different temperatures. Any one of these could seriously jeopardize astronomical work and should be avoided when locating observatory sites. The human eye also contributed to the difficulties of good seeing. Motes, described by Douglass as "undoubtedly . . . the remants of cells floating in the liquids which fill the parts of the eye-ball," interfered with vision and were often mistaken for planetary markings. Under certain conditions, especially when using high magnification, the outer surface of the eye cast shadows on the retina, also interfering with astronomical research.[13]

In several articles Douglass described his planetary work. He explained the high projections on Mars, often sighted thirty to sixty miles from the terminator, as cloud cover. The thin Martian atmosphere could cause these high clouds: "Of course cloud height could be greater on Mars as the atmosphere decreases in density on ascent much more slowly than with us." Commenting on the surface markings of Jupiter, Douglass concluded that two main currents moved the colorful bands of the planet's atmosphere, one moving the bands parallel to the equator and the other descending from the poles. This latter current (which NASA's Voyager I failed to observe in 1979) resulted from internal heat warming the atmosphere and causing it to rise over the poles and drift toward the equator.[14]

Later articles examined the mechanics of astronomical observations, reflecting Douglass's emphasis that an understanding of such phenomena was critical in locating observatories. He stressed the need to standardize conditions of observations. Deriving a reasonably simple and straightforward equation for the relation between telescope aperture, power of magnification, the planet's distance from the sun, albedo (reflecting power), and other data, Douglass constructed a table providing his readers with the proper combination of power and aperture for optimum observation. Other studies focused on atmospheric currents. Testing various sites in northern Arizona in February of 1899, while Boothroyd manipulated the telescope at the observatory, Douglass concluded that many of the currents that had disturbed work at the observatory were air waves from the nearby San Francisco Peaks. Using a twelve-inch mirror supplied by Clark and Sons, he checked sites at various distances from the mountains, finding increased waves as he approached the range. This article suggested the value of more isolated sites for future facilities and added new data to the growing awareness of the importance of site location for observatories.[15]

Douglass frequently had to defend the work at Lowell against the attacks of other astronomers. His first defense appeared in 1896 as a response to W.W. Campbell's review of Lowell's *Mars* (1895) in the *Publications* of the Astronomical Society of the Pacific. Campbell, on the staff of the Lick Observatory, challenged Lowell's initial assumptions and also questioned his use of observational data to support them. Composed by Lowell and Douglass, and published under Douglass's name, the observatory's rejoinder stated that Lick astronomers two years earlier had tried to discredit Lowell's work. The earlier attack, thought the Flagstaff scientists, was motivated by jealousy of Lowell's work as "too original." Now, the California scientists were attempting to "seize the credit of the results" and called the hypotheses "mostly old." Douglass and Lowell pointed to weaknesses in Campbell's review, but failed to counter the criticism.[16]

In his various writings, Douglass emphasized the superiority of the Flagstaff atmosphere for astronomical work. Only the Lick Observatory, the Harvard station in Peru, and the Flagstaff and Mexico sites of the Lowell Observatory had been chosen for their atmospheric qualities. He blamed poor seeing for the failure of other observatories to confirm the Lowell discoveries. At Lowell, the staff had frequently studied the markings on both Mars and Venus, with proper precautions to eliminate psychological influences or lens deficiencies.[17]

In one article, Douglass stated that every astronomer quickly adapts to the idiosyncracies of both his instrument and his site. Each knows the most favorable methods of making observations and he should be regarded as the highest authority on his own observations. This certainly was the case at Lowell, where the staff not only understood the site, but made observations with a consecutiveness which added weight to their conclusions. If other astronomers enjoyed the advantages of the Lowell site, he observed, they would confirm most, if not all, of the observatory's results. Concluding an article in the *Monthly Notices of the Royal Astronomical Society*, Douglass wrote heatedly:

> No matter how difficult to obtain, a just hearing is our right. No one is entitled to cry out against us until he can show that his atmosphere is approximately as good as the one through which Mr. Lowell discovered these markings [in this case, on Venus]. Let our dubious friends, who attempt to show that we as well as they are deluded, devote a portion of their valuable time to work at the telescope under better atmospheric conditions and no one will misunderstand the silence which will follow.[18]

By early 1900 Lowell began taking a more active role in the affairs of his observatory. During late spring, he and Douglass coordinated an elaborate plan to observe the solar eclipse of May 28. Lowell observed in North Africa and Douglass from Washington, Georgia (with the Harvard Expedition under W.H. Pickering), both stations along the line of totality. Lowell also sent Douglass a list of instructions for observatory work during the summer, including significant observations of Saturn. Douglass was to make detailed drawings of the planet, along with micrometric measurements of its diameter. Lowell also requested observations of Saturn's satellites, and instructed Douglass to examine Jupiter and Venus for markings and to determine the rotation period of Jupiter's several belts.[19]

The "Message from Mars" and the Break with Lowell

The next Martian opposition found Douglass again focusing on the red planet. Between August 11, 1900 and March 31, 1901, he concentrated on Martian observations, but gained little new information. On the night of December 7, however, Douglass saw a particularly bright projection near the Martian terminator. As was his practice, he telegraphed Lowell the next morning: "Last night projection north edge Icarium Mare lasted seventy minutes." Following his usual procedure, Lowell quickly forwarded this message to Harvard and European astronomers.[20]

The European press picked up this story and reported that Douglass had seen a message from the inhabitants of Mars. The civilization that had constructed the vast canal system had evidently attempted to communicate with its Earth neighbors. Immediately, American newspapers reprinted the story and Douglass received several telegrams and letters from interested laymen and reporters wishing more information. One letter proved especially disturbing to the observatory staff. Daniel E. Parks, a Denver attorney, requested details of Douglass's observations and offered a possible explanation:

> I observe from the papers that on December 7th last (1900) you observed a projection of some kind, on the planet Mars. Will you kindly advise me what the projection appeared to you to be, whether a dark material object, or a shaft of light? Suppose the people of Mars have built a monument ten miles square at its base and a hundred miles high, covered exteriorly with polished marble (which is the fact) would not the monument reflect a shaft of sunlight?

For the next several weeks, Douglass and other astronomers repeatedly explained that the observed phenomenon was nothing more than a cloud. By early February of 1901, public interest had finally subsided.[21]

Douglass's experience with the 1900 "message from Mars" illustrated the controversial nature of theories about Martian life. These theories had reached the public through Lowell's popular work *Mars* (1895) and H.G. Wells's classic science fiction novel *The War of the Worlds* (1898), both of which gained immediate wide popularity. Serious astronomers, however, were extremely skeptical of the existence of intelligent life on Mars.

Skepticism spread quickly after Lowell's book appeared. In 1895, George Ellery Hale, a distinguished astronomer and editor of the prestigious *Astrophysical Journal*, refused to publish further articles by Lowell because they appeared far too speculative. But the most persistent criticisms of Lowell's work came from the Lick Observatory. In his critical review, W.W. Campbell had emphasized that Lowell's conclusions agreed "perfectly" with those the Boston astronomer expressed before he began his observations. Campbell also questioned some of Lowell's later conclusions. Lowell had stated that along the broad equatorial belt of the planet, canals irrigated the deserts from both the north and south polar caps, regardless of which cap was closer. Campbell pointed out the improbability of this by constructing a parallel for Earth. Lowell's ideas would be similar to irrigating San Francisco, New York, Rome, and Tokyo from Antarctica, while using the North Pole to provide water for Chile and Australia.

Campbell also rejected the conclusion that terminator projections represented clouds, postulating instead that they were mountain chains parallel to the Martian equator lying across the terminator. Campbell criticized Lowell's reporting of data, especially the calculated distances from the planet's limb of terminator irregularities. Lowell reported these in seconds of arc to the third decimal point, while Campbell noted that all astronomers knew that such measurements often contained errors of more than a tenth of a second.[22]

E.E. Barnard joined the debate in 1897. He stated that the idealized drawings of Mars had led to many of the "queer and unnatural ideas" about the planet and its supposed canals. "If the marking [on a planet] is vague and uncertain it should be made so on the drawing and no definite outlines should be given where none exist," he said. W.H. Pickering's published account of the 1890–93 Arequipa work (for which Douglass had helped gather and reduce the data) also challenged Lowell's ideas. Pickering said that he had never seen the supposed double canals, and added that no matter how far away Mars moved during an opposition, the separation of double canals (as seen through the telescope by other observers) remained the same. Studying Schiaparelli's 1888 observations, Pickering attributed the phenomenon to an optical illusion:

> If merely a coincidence, it is certainly singular, that when the aperture of his telescope was doubled, the separation of the canals all over the planet happened to be reduced one half. It is also remarkable that the separation of the canals should happen to coincide so exactly with the observed separating power of the telescope he used.[23]

Douglass himself had long questioned the degree to which the "human equation" entered into astronomical observations. In late December of 1894, he had suggested to Lowell that more detail could be seen if he turned his attention "altogether to the light regions," indirectly suggesting a psychological explanation of some of the observed phenomena. In reply, Lowell recounted his own observations of the double canals, questioning whether they existed. "If they do not," he opined, "an interesting optico-psychic fact does." Both Douglass and Lowell, therefore, were aware of the psychological aspect of planetary observations. The astronomers took precautions to correct for this, publishing drawings all done by a single astronomer to minimize individual differences. They also restricted occupancy in the observatory dome to one astronomer at a time, hoping to prevent the power of suggestion from overcoming the power of observation.

Having doubted the reality of the double canals as early as 1894, Douglass recorded his glimpses of them during the 1898–99 opposition, but clearly emphasized their dubious character. This brought a sharp rebuke from Lowell in the *Annals*: "It is not possible to give with certainty a list of the double canals detected at this opposition," wrote Lowell in 1905, "owing to the hesitancy of Mr. Douglass to subscribe to his own delineations. He would draw canals double, and then doubt his drawings in his notes."[24] Lowell had evidently abandoned his earlier qualms about the reality of gemination, recognizing that the phenomenon added support to his theory of Martian civilization.

Douglass had become interested in work on "artificial planets" at the Lowell Observatory, hoping to investigate optical illusions resulting from lens inadequacies or atmospheric shortcomings. In December of 1898, he began making specific comparisons between the disks and planetary drawings, but the following year Lowell halted this exercise because, according to Douglass, "it was evident it cast doubt on some observatory publications." By the fall of 1900, however, Lowell had reconsidered, and he directed his assistant to continue the artificial planet work, focusing on the markings reported on Venus. He also suggested that a globe be placed at a suitable distance from the observatory and viewed through the telescope to see if lines drawn on the surface appeared to double. Douglass enthusiastically resumed his investigations, finding sufficient evidence to conclude that much of the surface detail seen on globes and disks was fictitious, strongly suggesting that canals on Mars and Venus might also be optical illusions. Lowell himself added to this possibility upon his return to Flagstaff in the spring of 1901. On his first observation of one of the disks, he drew a double line where only a shading existed.[25]

Throughout the early months of 1901, Douglass corresponded with many individuals about his work at the Lowell Observatory. He confided to his friend W.H. Pickering that the quality of the Flagstaff atmosphere had lessened over the years, and that he had started gathering meteorological records from other areas in Arizona, hoping to find a more suitable observatory site. Douglass wrote to Dr. Joseph Jastrow, a psychologist in Madison, Wisconsin, and asked about the human eye's ability to perceive and interpret details similar to those supposedly seen on planetary surfaces. What effect did others' drawings have on an individual's perception? What effect did certain colors and intensities of light have on the perception of such detail? Douglass continued to consider these questions for many years.[26]

Douglass was becoming increasingly disenchanted with Percival Lowell and his astronomical methods. Writing to William H. Pickering in early March of 1901, he stated, "It appears to me that Mr. Lowell has a strong literary instinct and no scientific instinct." He felt that his employer had been unfair in expressing great disappointment when Douglass failed to complete the second volume of the *Annals* in four months. Lowell's inconsistent stand on the artificial planet work also gave Douglass pause, as did his superior's continued insistence on the possibility of canal systems on Venus. Douglass was concerned about his own career. He mentioned Lowell's "strong personality, consisting chiefly of immensely strong convictions," which had led Douglass to support many ideas he now regretted. He asked Pickering if there existed any chance for a position at an observatory in South America, emphasizing his desire to work again with him.

Pickering replied that Lowell's work was generally held in low regard by the astronomical profession. Douglass's comments on Venus pleased Pickering, who had been disappointed by his young friend's earlier support of Lowell's controversial ideas. He agreed that Lowell had behaved unreasonably in expecting him to finish the *Annals* in four months, but could not offer Douglass a position at an observatory. He advised him to remain with the Lowell Observatory, "unless the loneliness is too oppressive."[27]

Douglass recognized that the observatory could salvage its reputation if it could divorce its scientific researches from Lowell's literary speculations. To this end, he wrote Putnam a long letter in March of 1901, pointing out the shortcomings in Lowell's work which had caused professional astronomers to dismiss him. He asked Putnam to urge Lowell to make his work more scientific in character:

> His work is not credited among astronomers because he devotes his energy to hunting up a few facts in support of some speculation instead of perseveringly hunting innumerable facts and then limiting himself to publishing the unavoidable conclusions, as all scientists of good standing do, in whatever line of work they may be engaged.

Putnam should convince Lowell to go more slowly in his scientific writings and focus on his popular contributions, clearly his greatest ability. This would be the best course of action, for, in Douglass's words, "I fear it will not be possible to turn him [Lowell] into a scientific man." After his signature, Douglass added the postscript, "Please consider this letter as between ourselves only."[28]

Putnam kept Douglass's letter confidential for a few months, but soon showed it to Lowell. In July of 1901, Lowell dismissed Douglass from the Lowell Observatory without explanation. Throughout the late summer, Douglass attempted to gain reinstatement, arguing that he should be entitled to the spectroscopic study Lowell had promised him and emphasizing that he needed the position to complete a great deal of unfinished work. Lowell refused his request: "I have at present all the assistants I want at Flagstaff."

Judge and Teacher

Douglass's dismissal from the Lowell Observatory was a traumatic shock. For seventeen years he had been actively engaged in astronomy, both as a student and as a researcher. Now, he found himself divorced from the career for which he had been trained. In late August of 1901, Douglass wrote to W.W. Campbell of the Lick Observatory that "Mr. Lowell and I have had the break-up that I have thought possible since the summer of [18]96," and asked if there existed any positions at the California observatory which he could fill. Campbell could not assist his younger colleague. Douglass also attempted to rejoin the Harvard College Observatory and to gain an appointment at the U.S. Naval Observatory, but no positions were open at either facility.[30] Unsuccessful in his attempt to secure astronomical employment, Douglass spent several months traveling in northern Arizona, living off his accumulated savings. He also invested in local mining ventures during 1902, and employed his knowledge of chemistry to open an assay office in Flagstaff, a venture which proved successful for several years.[31]

As a resident of Flagstaff for eight years, Douglass had made many friends. This wide circle of acquaintances led him to consider entering local politics as a temporary career. Even though he still considered himself an astronomer, Douglass's economic needs had to be met. The salary from a local political office seemed a good means to this end.

During the late summer of 1902, therefore, he announced his candidacy for the Republican nomination for probate judge. Despite his lack of legal training and experience, neither of which were formal requirements for the office, Douglass easily gained the party's nomination in the convention held on September 22. The Flagstaff *Coconino Sun*, a Republican newspaper, enthusiastically supported Douglass, emphasizing his "good judgment and sterling honesty." The *Sun* also focused on the nominee's personal characteristics:

> Mr. Douglas [*sic*] has been a resident of Flagstaff for the past eight or ten years; is a graduate of Harvard University, and a native of Massachusetts. With all these advantages there isn't a cowpuncher or lumberman in the whole section with whom he is ashamed to rub shoulders. He is here because he likes people and the country, and there is no question but those who know him heartily reciprocate the liking....[32]

Douglass ran a very personal campaign, traveling throughout Coconino County to introduce himself and talk to the voters. His warmth and charm easily won friends and support at every stop. The *Sun* reported: "So far he has met with splendid success and all hands, from logging camp to the sheep and cattle camps, are his staunch friends. They can't help it when they know him." Douglass's friendliness, popularity, honesty and intelligence were emphasized again and again in supporting newspaper articles. After a busy campaign, Douglass defeated S.S. Acker, his Democratic opponent, by a vote of 519 to 470. The two candidates split the outlying districts in Coconino County, but Douglass carried both Flagstaff and Williams, the only population centers in the county.[33] Over the next two years, Douglass proved himself an able probate judge and made an excellent impression on his political associates. He served on several Republican committees and easily gained reelection in 1904.[34]

While serving as probate judge, Douglass diligently pursued his interest in astronomy. He made several observing trips to the small Mt. Lowe Observatory above Altadena, California, during the spring of 1903, observing Mars during its opposition.[35] Douglass's other attempts at meaningful astronomical pursuits, unfortunately, proved disappointing. He was unable to find any observatory vacancies during his judgeship, and a proposal by Harvard Observatory to publish some of Douglass's Flagstaff work met the strong opposition of Percival Lowell.[36]

The summer of 1905, however, proved rewarding on a personal level. On July 1, the thirty-eight-year-old Douglass left Flagstaff on an extended trip to Los Angeles, where he and his fiancée of several years were married. Ida E. Whittington had come to Flagstaff a few years earlier to teach music privately. By 1904 she was instructing in the local schools and serving as assistant principal. Nine years younger than Douglass, Ida had been born in Baltimore but grew up in Kansas City and Los Angeles, where her father was a gentlemen's tailor. Andrew and Ida had contemplated marriage since 1902, but did not announce their engagement until June of 1905. The marriage ceremony took place on August 3 in St. Paul's Cathedral (Episcopal) in Los Angeles. After a brief

California honeymoon in Venice and San Diego, the couple returned to Flagstaff on 11 August.[37]

After his return, Douglass had a few brief weeks to prepare himself to teach at Northern Arizona Normal School. The previous March he had taught Spanish at the school, being the only qualified person available, and was soon offered a full-time position as instructor in Spanish, history and several other courses at a salary of $1200. To prepare himself for the new duty, Douglass went to Berkeley and took a two-week education course at the University of California's summer session.[38]

Before he began teaching, he wrote to Arizona Attorney General E.S. Clark to confirm that he could legally teach at the Normal School while serving as probate judge. The answer was somewhat disturbing. Clark said that he had "many serious doubts" concerning the legality of Douglass serving in both capacities, but added that he would check more closely into the matter. In a hurried response, Douglass emphasized that during the preceding spring his work as probate judge had in no way suffered as a result of his teaching activities. Although retaining doubts, Clark later advised that in the absence of a law pertinent to the matter, Douglass could accept the teaching position. Henry Fountain Ashurst, Coconino County district attorney and later a distinguished senator, also assured his friend Douglass that he would be violating neither the spirit nor the letter of the law by holding both positions. For the next nine months, Douglass served as one of six instructors at the Normal School.

Douglass's entry into the teaching profession proved to be the path of his return to meaningful science. By the end of the 1905–1906 academic year, Douglass had secured a position at the University of Arizona in Tucson. He had previously applied to two teachers' agencies in his search for academic employment and had also advanced his name for the chairmanship of the mathematics department at Kenyon College in Ohio. None of these actions proved successful. F.S. Luther, Douglass's mentor at Trinity College, apologized for not being able to offer him a vacancy in astronomy, explaining that the position had been filled a few months earlier.

On May 12, 1906, the Flagstaff *Sun* announced Douglass's new appointment. He resigned from the Normal School a few weeks later, but remained probate judge during the early part of the summer. Douglass ended his twelve years in Flagstaff as a well-liked and respected member of the community, but his future lay in Tucson, far to the south. For Douglass, however, his new position represented far more than a change in climate and scenery. After five years of uncertainty, he saw in southern Arizona an opportunity to return to the astronomical world from which he had been isolated.[40]

4

The Steward Observatory

SOON AFTER HIS ARRIVAL in Tucson in 1906 to assume his position at the University of Arizona, Douglass realized that astronomy could be a major interest in the area. Despite the poor weather conditions that had plagued his initial observations twelve years earlier, he became convinced after several months of more detailed tests that southern Arizona presented an outstanding site for a significant observatory.[1] During the next two decades, Douglass focused his energies toward building such a facility.

In a public lecture given on campus the night of February 13, 1907, Douglass stressed the advantages Tuscon offered to astronomy. Reported in the next day's newspaper, this lecture marked the first of his appeals on behalf of astronomy in southern Arizona. His most energetic attempt came a few months later when he wrote a long letter to the appropriations committee of the territorial legislature, requesting that it increase the university's funding from $50,000 to $60,000, the extra sum to be used to equip an observatory.[2] Emphasizing the clear Arizona air, he stated that a southern Arizona site would be much better than one in California, and might possibly be the best site in the nation. He also appealed to territorial pride, informing the committee that virtually every well-furnished university possessed a telescope of eight to sixteen inches aperture, while the Tucson institution had nothing more than a second-hand four-inch instrument with several parts missing. Douglass added that many students had expressed interest in taking an astronomy course, and one had left Tucson to attend a California school which offered such instruction. Despite his plea, the committee ignored the professor's request as too expensive for the territorial budget.

University duties prevented Douglass from actively pursuing his observatory campaign until April 1908, when he wrote a major article for the Tucson *Star*. Rather than focus on the scientific research which could be accomplished by a local observatory, he stressed the value of such an installation to the territory itself. A large telescope would advertise Pima County, underscoring its clear and steady air. In addition, the observatory would be a great addition to the territory's educational equipment. Playing on the growing desire for statehood, Douglass closed his article by stressing that the establishment of an astronomical facility would show that the citizens of Arizona recognized their many advantages. "Statehood will be safe," he wrote, "in the hands of such public spirited citizens."[3]

The future of the territory was also uppermost in the mind of George W.P. Hunt, an influential territorial senator who shared a train journey with Douglass from the East during the late summer of 1908. Writing in December, Hunt expressed hope that Douglass had not forgotten their discussion about bringing an observatory to southern Arizona. Hunt wished success in securing funds for such an installation, which would benefit both science and the territory. He closed with an offer of any assistance possible and the wish of talking to Douglass again. On December 31, Douglass took advantage of Hunt's offer and asked his advice on the possibility of legislative assistance. He suggested that wealthy residents of the territory might be interested in financing an observatory, if the legislature would meet them halfway. Douglass asked Hunt if this seemed a good idea and if the legislature would respond favorably and add a significant amount to the regular university funding. Of equal concern, he questioned whether the failure to secure

$20,000 to $40,000 at present would prejudice the legislature and private patrons against further attempts. Once again, however, the project failed to advance beyond speculation.[4]

Douglass continued his efforts. In a few months, he penned another article for the *Star*, again emphasizing the value of an observatory. On February 12, 1910, he wrote to the philanthropist Mrs. Russell Sage, unsuccessfully asking for her assistance in raising the estimated $50,000 necessary to construct an observatory. A letter to regent Merrill P. Freeman, who in 1891 had bought the small telescope used by the university, also failed to stir interest, even though Douglass stressed the national and local benefit of such an installation.[5]

By December of 1911 Douglass had developed a more elaborate program, which he outlined in a letter to University President A.H. Wilde. He was collecting various letters from astronomers, legislators, and other territorial officials regarding the value of an observatory in southern Arizona and the assistance available in the territory for financing and maintaining such an installation. Estimating that the desired instrument would cost $50,000, Douglass hoped to take his collection of letters and approach persons of wealth within the territory and those outside. He asked President Wilde for a letter stating his views on such a project.[6]

Wilde wrote that while Douglass's plan was interesting, he doubted whether the university should mount a campaign for a $50,000 instrument. After all, he said, an auditorium, museum, and agriculture building presented "immediate and imperative needs," while an observatory remained somewhat of a luxury. If wealthy Arizonans were to be approached, they should be asked to support the more important buildings first, before investing in an instrument that would cost as much as a building. Wilde closed by expressing his appreciation for Douglass's suggestions and enthusiasm, but added, "the comparatively small number of students who would take the work and the great service that would come to us in the other investments would lead me to prefer the latter to the telescope at this time."[7]

Outside the territory, sentiments were different. In March of 1908, W.H. Pickering wrote Douglass that he wished his young friend had a good telescope in Tucson, lamenting, "Certainly there ought to be one well located telescope in the United

States with a reliable observer behind it." Within a few months, Pickering and his brother, the director of the Harvard Observatory, arranged for Douglass to borrow a medium-sized lens which had been part of an instrument no longer used at Harvard. While not the major observatory the Tucson astronomer desired, this eight-inch telescope and mounting was a vast improvement over the university's existing equipment. The university agreed to pay the costs of transportation and provide a suitable housing for the instrument.[8]

Delays in Cambridge prevented the early completion of a telescope tube for the lens, although it had been promised for late fall. The University of Arizona, however, carried out its role. A new science hall was being constructed at the time, and President Kendrick C. Babcock authorized that an observatory room be included on the top floor. Not wanting a dome in so conspicuous a place, he insisted on installing a sliding roof over the room, which proved serviceable, if somewhat unorthodox. Construction of the telescope tube in Cambridge proceeded slowly because of other demands on the observatory staff's time, but the mounting reached completion in early March of 1909 and was shipped to Tucson.[9]

Following the instructions sent by Harvard, Douglass mounted the telescope during the first week in April and tested the instrument on the ninth. Tracing stars accurately required a clock drive, so Douglass began designing the needed machinery. Built by the mechanical engineering department, the clock ran adequately with periodic refinements and adjustments. The following year, Douglass added a wireless receiver to the astronomy equipment to obtain time broadcasts from government stations on the west coast.[10]

With this larger telescope, Douglass could conduct original research which had been impossible for the past seven years. The new instrument enabled him to continue his investigation of the atmospheric quality of the Tucson area, confirming his favorable evaluation of the region's seeing. The first significant original contribution came in early 1910, when Douglass observed an unusual phenomenon associated with Comet *a*, the first comet observed that year. While the cometary tail grew in length from eight degrees to forty-five degrees, a normal occurrence, it also branched or forked.

Douglass also initiated observations of Jupiter, but the most exciting event of the year was the return of Halley's Comet. In addition to reassuring the public that the earth was in no danger from passing through the comet's tail, Douglass recorded an apparent splitting of the comet's nucleus. He noted that the two pieces appeared to be thirty seconds apart with a fine line connecting them. The next night only a faint nucleus remained visible, which decreased until it vanished on 28 May.[11]

During the next few years, Douglass found the telescope extremely valuable. He continued to study Jupiter's satellites, meteors, and other astronomical phenomena, and also used the telescope as a teaching tool, assembling one astronomy class at 4:00 A.M. to observe the Leonid meteor shower and Mars. On November 8, 1914, he involved the astronomical equipment of the University of Arizona in community service, assisting the *Citizen* in announcing the approach of a group of racing cars engaged in a long-distance race.[12]

Significant astrophotographic work also occupied Douglass's time. He continued studying the zodiacal light, as he had done at the Lowell Observatory, and perfected his technique in preparing exposures of the phenomenon. In mid-1915, he submitted a collection of photographs to the London exhibition of the Royal Photographic Society. The organizer of the exhibition was so impressed with the description of Douglass's work involving long exposures of faint objects that he requested permision to publish it in their journal. Presented before the Royal Photographic Society on December 14, 1915, Douglass's contribution appeared in the February 1916 issue of *The Photographic Journal.*[13]

Although the Harvard eight-inch refractor enabled Douglass to carry on an impressive amount of work, it lacked the potential for the great contributions of an established observatory. Douglass continued to prepare estimates for a telescope and to search for possible financial assistance to build it. Upon receiving price quotations from Pittsburgh's long-established astronomical firm of Brashear, he drew up a comprehensive plan for a university observatory. The cost of the telescope alone ranged from $34,000 for a thirty-six-inch reflecting instrument to somewhat less than $47,000 for a twenty-six-inch refractor, slightly larger than that at the Lowell

Observatory in Flagstaff. Expenses for freight, local construction, a building for the astronomy department, and necessary accessories added as much as $37,000 to the estimate. Douglass also calculated that $4000 per year would be required for maintenance.[14]

Douglass's attempt to secure financial backing for the project proved extremely frustrating. During the summer of 1914, he tried unsuccessfully to interest Arizona's mining interests in financing an observatory. He also urged state officials to consider the value of an astronomical installation in Tucson. Emphasizing that Tucson presented a superior site to Flagstaff and that Percival Lowell's work represented poor science, Douglass urged Tucson state senator John T. Hughes to support an observatory which "would be a drawing card for both graduate and undergraduate students and would attract tourist travel through our section." Hughes, who had previously assisted the university in obtaining its new agriculture building, assured Douglass that if the university urged such an appropriation, he would do everything possible to see it through the legislature.[15]

Douglass again approached regent M.P. Freeman about the facility. In an eleven-page exposition on Tucson as an observatory site, he recounted the clarity of the night sky in Tucson and the astronomical contributions already made, such as the unique observations of the nuclear division of Halley's Comet four years earlier. Local authorities had suggested that the next legislature, meeting the following February, would probably appropriate money for the observatory if an "entering wedge" could be supplied through a gift, especially if that gift remained contingent upon matching funds from the state. Asking for Freeman's assistance in securing a contribution of some $25,000, Douglass assured his friend that anyone giving such a sum could certainly expect to have the observatory named after him.[16]

Later in the summer, Douglass began enlisting the aid of his professional colleagues. Writing to such eminent astronomers as George Ellery Hale, E.B. Frost (director of the Yerkes Observatory), W.W. Campbell, and W.H. Pickering, he asked for letters giving their opinions on the desirability of locating a major observatory in the Tucson area. Letters from such nationally recognized scientists would, in his view, help to convince the state legislature to appropriate the needed funds. He also asked for any information on the possibility of securing the loan of a telescope or the cooperation of a respected observatory in establishing the Tucson installation. As always, he remained open to suggestions for sources of an initial gift to encourage legislative action.[17]

Douglass brought other forces to bear in his attempt to secure an observatory. Reminding Governor George W.P. Hunt of their earlier discussion in 1908, Douglass sent him a three-page proposal which the governor promised to place before the Board of Regents himself. Douglass later asked Philip Fox, astronomy professor at Northwestern University, to help secure the assistance of the American Astronomical Society for the observatory. On August 28, the society adopted an appropriate resolution supporting Douglass's plan. University President R.B. von KleinSmid supported the observatory in appearances before community organizations, and the Tucson press urged the legislature to fund the proposed astronomical installation.[18]

By October of 1914, however, the project's future looked bleak. Unable to attract private contributions, supporters of the observatory were forced to rely on legislative appropriations. Unfortunately, a new School of Mines building had been requested at the same time. As M.P. Freeman cogently argued in a letter to Douglass, it was unlikely that both a mining school and an observatory would be funded, and the former would be far more popular. "So you see, Doctor," concluded Freeman, "we are confronted with a condition that is going to work against the Observatory, possibly none too popular anyway with the class of men that we select to make our laws."[19] When the legislators met, Freeman's prediction was confirmed. Requesting $84,350 to build and equip the observatory, the University of Arizona budget emerged from the appropriation committees' hearings intact, but the House and Senate considered the observatory a dispensable luxury. The legislature of 1915 adjourned that spring with Douglass's observatory still the dream of a frustrated astronomer.[20]

The Observatory Finds a Benefactor

Hopes for the project finally brightened in the fall of 1916. On October 18 President von KleinSmid officially announced that an anonymous benefactor had given $60,000 to the university to build an observatory. He also stated that various eastern mining firms which had interests in Arizona had contributed $75,000 for a new mining building. After announcing the gifts, von KleinSmid dismissed classes at the university for the rest of the day. Various celebrations took place on campus during the afternoon and evening.[21]

The anonymous gift came from Mrs. Lavinia Steward, who had moved to Tucson from Joliet, Illinois, in 1898. She and her husband soon resettled at the little town of Oracle, thirty miles northeast of Tucson, because of its higher elevation and cooler temperature, and remained there for the rest of their lives. By 1916, Mrs. Steward had expressed a desire to do something for the University of Arizona in memory of her husband, who had died twelve years earlier. An amateur astronomer, Mrs. Steward proved quite sympathetic to von KleinSmid's suggestion that she help finance an observatory on campus. That summer, she gave $60,000 to establish what would soon be known as the Steward Observatory.[22]

Plans for the observatory now began in earnest. Douglass quickly organized the information he had collected into a report, which he presented to von KleinSmid on November 10. A thirty-six-inch reflector, mounting, dome, freight, and construction, he said, would cost about $46,000. With the necessary building and accessories, he estimated that the total would be $57,700. Because of the temperature extremes in Arizona, especially during the summer, Douglass recommended that the dome be built entirely of steel with double walls to improve insulation. The dome might have to be cooled in summer to keep the daytime temperature close to that at night, so that the mirror would not be adversely affected by temperature fluctuations. Regarding the mounting of the telescope, Douglass advised that Warner and Swasey, the Cleveland engineering firm which had constructed the mounting for the Lick thirty-six-inch refractor, was very busy filling war orders. Although

A.E. Douglass at the time of the Steward gift to the University of Arizona.

this situation could hamper design work on the Steward mounting, the firm had promised to return to their telescope contracts as soon as possible.[23]

The next step was to select a site for the new observatory. Douglass considered several, but he soon narrowed the choice to two locations: the hills three miles west of the university, or the campus itself. Although the seeing on the hills was slightly better, the problems of locating on the most advantageous hill far outweighed the slight improvement in atmospheric quality. A location on campus also presented disadvantages. The western part of the campus contained the principal university buildings, rested near a major city street, and possessed the lowest elevation on campus. To compensate for these shortcomings, an observatory would have to be built so that the floor of the dome rested fifty feet above the ground level, an expensive and inconvenient proposal. The east end of campus proved far superior because of a twenty-five-foot ridge and its isolation from light interference and heat-producing buildings, both detrimental to astronomical research. Douglass therefore recommended to the university president that the Steward Observatory be erected on the eastern site, with safeguards against future tall buildings. He also proposed that lighting systems in the area be placed under the control of the astronomer in charge of the facility. Insulated from bright lights and large buildings, the observatory could anticipate a long and useful career.[24]

The heart of the new telescope would be its large mirror, and Douglass overlooked nothing in his search for the finest available. The Carnegie shops in Pasadena, currently working on the 100-inch Mt. Wilson reflector, indicated no interest in making a mirror, but the project's chief optician sent detailed test instructions and suggestions for Douglass to use on whatever mirror he could obtain. With this information the Steward telescope could meet the same standards as any in southern California.[15]

Douglass next turned his attention to obtaining a glass disk from which the mirror would be made. The finest disks came from the St. Gobain factory in France, but because of the war this source was no longer available. Douglass therefore placed an order with James B. McDowell of the Brashear Company in Pittsburgh, with the understanding that manufacture of the glass would take no more than five months. He stressed that the glass quality must be equal to that of earlier reflectors, and asked when he would need to furnish exact specifications on the mirror.[26]

Simultaneously, Douglass negotiated with Warner and Swasey for a mounting and dome. A great deal of money could be saved, he said, if the firm constructed the pier out of concrete or cast iron instead of structural steel. Because of the warm southern Arizona climate, he also suggested that a lightweight dome be made. By saving money on the mounting and dome, he hoped that he could order a larger mirror, but the Warner and Swasey estimates quickly showed the improbability of this plan. The mounting would cost at least $33,000, with the dome adding almost $11,000 more.[27] By the end of December, Douglass received from the Cleveland firm a set of preliminary designs and estimates for use in deciding the details of the Steward Observatory.

To add to his knowledge of telescope mirrors, Douglass sought advice from many astronomers, included George Ellery Hale of the Mt. Wilson Observatory. Hale allowed Douglass to observe through the sixty-inch Mt. Wilson reflector in late February of 1917. His observations, especially those of Mars, confirmed his decision that a reflecting telescope would prove the most suitable for his planetary studies.[28] It would also be cheaper and easier to build than a refracting telescope such as Douglass had used at the Lowell Observatory.

Obtaining the Steward mirror proved exceptionally difficult. The Nation Optical Glass Company, to which Brashear had given the contract to cast the mirror blank, had just installed a new annealing oven at its glass factory. The annealing, or curing, of the glass was the crucial stage in the process and could not be rushed. By mid-March, the thirty-seven-inch mirror had been cast and put in the annealing oven, where it would stay until early April. A serious setback occurred on April 2, when the disk was taken out of the oven and found to be badly cracked.[29]

During the following four months, Douglass tried to make other arrangements for a suitable piece of high quality glass. Corning Glass Works, later to be famous as the manufacturer of the 200-inch mirror for Mount Palomar, could do none of the work Douglass requested. Pittsburgh Plate Glass Company replied that they could furnish glass in the necessary size, but they had no experience in the complicated annealing process. The firm offered to work at cost if Douglass or Brashear wanted to experiment, but both Brashear and University of Arizona officials opposed any expenditure for another organization's experiments.[30]

Meanwhile, plans for the telescope machinery advanced more smoothly. On February 4, 1917, Douglass telegraphed an order to the Cleveland engineering firm of Warner and Swasey for the body and mounting of the telescope at the quoted price of $33,600. Before signing the contract, he wanted to make sure that Warner and Swasey would insure the instrument until delivery. The company was located in a German area of Cleveland which offered a prime target for supposed saboteurs. Once this assurance reached Tucson, university officials signed the contract on March 19. Douglass agreed to make at least two trips to inspect the progress of construction, the first to take place following completion of the preliminary drawings, estimated for mid-April. The university arranged to make quarterly payments, with equal amounts to be paid upon completion of the drawings, after the mounting was half-completed, upon completion of the mounting, and at the time of installation in Tucson.[31]

The war quickly intruded into Douglass's plans. Journeying to Cleveland in late June to confer with the contractors, Douglass urged Warner and Swasey to expedite their work. Mrs. Steward had developed tuberculosis and might not last another year. Obviously, completion of the observatory would mean a great deal to her. In addition, Mars would be in a very favorable position in the spring of 1918, providing an excellent opportunity to use the new instrument. Finally, wrote Douglass, the Pacific Coast Branch of the American Astronomical Association was planning its April 1919 meeting for Tucson. A new observatory would be an attractive addition to that program. Warner and Swasey promised to move ahead on the telescope mounting, even though the firm held a $2,000,000 war contract.[32] This promise proved impossible to keep by late July, when Warner and Swasey had to abandon completely the Steward project because of their military commitments. Although Douglass tried to urge a resumption of work on the mounting, the federal government refused to allow such civilian production. Lavinia Steward died in August, her observatory little more than a collection of plans and proposals.[33]

Despite the war, Douglass wrote to James McDowell at Brashear on March 1, 1918, asking if the optical company could "take a shot" at the thirty-seven-inch mirror sometime during the summer. Because the firm was completely engaged in war orders, this proved impossible. Douglass also wrote to W.R. Warner, president of Warner and Swasey, asking much the same question:

> I am writing my annual letter to see if it isn't some way possible for you to have some progress on the plans of the reflector ready for me this summer. I hope to make a trip east and I shall come through Cleveland to see if something can't be accomplished.
>
> I know the urgency of the war work and sympathize with it thoroughly, but isn't it possible to get this thing some way by the good management for which you are famous? Astronomical opportunities are moving by that will not return in my lifetime.

Warner answered that no possibility existed to begin work on the telescope contract, because government inspectors were in the plant watching every move. Nothing must detract from the war effort.[34]

When word came of the war's end, Douglass leaped into action. At 1:30 A.M. on November 11, 1918, within hours of the armistice, he telegraphed Warner and Swasey to "Wire me when you can resume work on telescope." L.B. Stauffer, the firm's secretary, answered in a letter the following day, writing that there had been no change in any of the government contracts. Furthermore, the plant manager had just left for a long-needed vacation in California, and no work on the telescope could possibly begin until his expected return in mid-December. Douglass also wrote to the Brashear company, asking when they could begin mirror work. McDowell replied that Brashear was very busy, and he expected a navy contract for sextant mirrors to last for "a couple of years." Until these conditions changed, McDowell said that they could undertake none of their regular line of work.[35]

By mid-February 1919, Warner and Swasey notified Douglass that the firm planned to resume work on the mounting for the Steward Observatory. The design staff completed preliminary drawings by early April, and asked Douglass to come to Cleveland for a conference. Douglass left Tucson on April 10 and spent several days consulting with both Warner and Swasey in Cleveland and Brashear in Pittsburgh. Douglass and the contractors coordinated their activities so that the mounting, mirror, and building specifications would complement each other.[36]

The Cleveland firm spent the rest of the year working on final plans for the observatory. By mid-July of 1919, the firm requested precise dimensions of the base and clock drive from Douglass. In early September, Douglass traveled to Cleveland and approved the plans. Pursuant to the contract, this approval marked the due date of the first one-quarter payment of $8400, which Douglass requested of President von KleinSmid. After a slight delay while the Arizona Board of Regents considered the wisdom of partial payments, a check for the amount reached Warner and Swasey on 5 November.[37]

Douglass expressed some concern regarding the secondary mirrors for the telescope. These small mirrors reflected the light from the main mirror to the eyepiece of the instrument at the focus. In the original plans, Douglass had called for octagonal mirrors because they weighed less and cut off less light than the traditional round secondaries. Grinding a mirror to a polygonal shape, however, could easily distort the mirror, reducing its value for accurate astronomical work. J.B. McDowell, president of Brashear, advised Douglass that circular mirrors would be the best, causing Douglass to ask Warner and Swasey for an estimate of the time and cost to redesign their plans. Warner Seely, the Cleveland firm's new secretary, replied that circular mirrors would increase the mirror assemblies' weight by fifty percent over the octagonal ones, requiring rebalancing of the entire instrument. Douglass conveyed his original preference for polygonal mirrors to Brashear. For the moment, Douglass's ideas carried the day.[38]

The manufacture of the primary mirror, unfortunately, remained mired in a pool of difficulties. By mid-April of 1919, Brashear had arranged for the Spencer Lens Company of Buffalo, New York, to cast the thirty-seven-inch disk, after which Brashear would do the optical work at a total cost of $6000.[39] The

Spencer Company, like all other American glass companies, had never attempted to make a large glass disk for a telescope mirror, all such glass having previously come from European sources such as St. Gobain. The company was convinced, however, that by finding a piece of crown glass of the correct size, they could produce a mirror blank for Douglass, using techniques shown to be effective on small, experimental mirrors. McDowell reported the Spencer news to Douglass, admitting that he remained puzzled as to why Spencer did not attempt to pour molten glass into a mold of the proper size.[40]

By early autumn, both Douglass and McDowell questioned the possibility of Spencer's success. The Brashear company had heard nothing from the Buffalo glass-makers since July, and final correction of the optical accessories made by Brashear depended on completion of the primary mirror. In reply to Douglass's inquiry, Spencer manager H.N. Ott wrote on October 15 that they had still not found a suitably large piece of glass to begin work on the mirror. Ott concluded with the frustrating statement: "If we do not find something before long we shall have to make some arrangement for pouring a pot of glass into the form you desire." Pursuing another source, McDowell had written to the French firm of St. Gobain, which had cast telescope mirrors before the war, asking if they could supply an adequate disk. Their reply brought the disheartening news that the factory had been destroyed during the war. Douglass and McDowell continued to urge the Spencer staff to make some sort of delivery promise so that work on the telescope could be continued, but with no success.[41]

As the year 1919 came to a close, Douglass faced total frustration with the Steward Observatory. At every turn for the past three years his plans for the swift completion of a major astronomical facility had been thwarted by numerous difficulties, accidents, and misfortunes. The Steward bequest in 1916 had been one of the great moments in Douglass's life, but his plans and ideas for the installation had been denied any expression in physical reality. The Tucson astronomer, however, focused his gaze on the ultimate completion of "his" observatory, confident that it would make lasting contributions to astronomy and to the University of Arizona.

5

Astronomy in the Desert

THE BEGINNING OF A NEW DECADE brought little relief from Douglass's frustration over delays in the establishment of the Steward Observatory. The first weeks of 1920 found him fretting again about the dome. Corresponding with the Chicago architecture firm of Schmidt, Garden, and Martin, he collected several proposals ranging in cost from $2180 for a wooden dome to $6500 for a steel structure. By late March, Douglass had decided that a wooden dome with a curtain-like shutter mechanism would best satisfy his requirements. The potential variations in the southern Arizona climate caused some concern that the dome would change shape and become difficult to move on the sill, leading Douglass to insist that some provision be made so the wheels on the dome base would move if changes in shape occurred. Douglass soon received a complete estimate for building the Steward dome. The total cost would be over $3000, significantly above the expense of local construction. In late May, Douglass wired the Schmidt firm that he had decided to have all work done in Tucson.[1]

Douglass turned again to Godfrey Sykes, now of Tucson, who had built much of the Lowell Observatory. Sykes designed the Steward dome to be built of a canvas-covered steel frame.

The frame was constructed by Tucson Iron Works and the building itself by Lyman and Place. The construction went well and by fall the dome and building were ready for occupation.[2] Only the telescope remained to be completed.

Warner and Swasey continued construction of the telescope mount during 1920, with Douglass visiting their plant in mid-September to check on progress. The mounting plans had developed well, with plans of the building, dome, and telescope exchanged throughout the summer and fall to insure that the observatory would be accurately constructed. Douglass again urged a speedy completion of the work, as he planned for the Southwest Division of the American Association for the Advancement of Science to meet in Tucson in December of 1921. The Cleveland engineers promised to make progress, but they had to obtain custom-made ball bearings from Sweden, the only source of sufficiently high-grade steel. Another year should see the mounting completed.[3]

Building the Steward Telescope

The large mirror for the telescope caused even more difficulty. Having heard nothing from Spencer Lens Company, Douglass wrote the firm in January of 1920, saying that he was "getting very anxious" about the telescope disk. H.N. Ott, the Spencer manager, informed Douglass in late January that they had not been able to make a suitable mirror blank. The firm's experts had several ideas, but had to confess the experimental nature of their work on the Steward glass. "We may succeed," wrote Ott, "and again we may not." He told Douglass that if the glass could be obtained in some other way, the observatory should take advantage of the new offer. "In all fairness," he concluded, "we cannot ask you to wait longer because of our inexperience."[4] Unable to find another source for the glass, Douglas had no choice but to wait.

In May, events took an encouraging turn. The Spencer Company had been moderately successful in making large disks, only to encounter trouble controlling the gas-heated furnaces used for annealing. To deal with this problem, the firm had called in experts from General Electric to design and construct a special annealing oven to be electrically heated. The furnace was also to have automatic temperature regulators to insure the proper cooling rate for the glass. General Electric had promised delivery in June, but the complicated design details postponed completion until late fall. Spencer cast the Steward disk in December and placed it in their new annealing oven.[5]

While waiting for the results of Spencer's latest effort, Douglass had to concern himself with the telescope's secondary mirrors. Brashear could not construct the polygonal mirrors desired by Douglass, forcing Warner and Swasey to halt its drawings. If other mirrors were to be used, the firm needed to know their exact size and shape. By mid-January of 1921, Douglass had all but given up on polygonal mirrors, and asked J.B. McDowell if he did not think "that this is the kind of error which you should have protected me from." The Brashear president wrote, "We knew that a proper surface could not be obtained on a rectangle of the dimensions given." On receiving this first clear statement of the difficulties involved, Douglass made a marginal note on the letter, "why didn't you tell me." By early May, Warner and Swasey had conferred with Brashear, and both firms advocated the use of oval mirrors, a decision Douglass reluctantly accepted.[6] The shape of the secondary mirrors proved to be the last major hurdle to completing the Steward mounting. With the arrival in Cleveland a few weeks later of the finder scope, the original four-inch refractor belonging to the university, Warner and Swasey were able to make the telescope ready for Douglass's inspection in June.[7]

Douglass arrived in Cleveland on June 8 to inspect the telescope mounting at the Warner and Swasey plant. For the next few days, he made as many tests as possible, during both day and night, concluding by the eleventh that the "Instrument is fine."[8] Warner and Swasey began dismantling the telescope for shipment after Douglass left in mid-June. By July 9, the dismantled mounting, weighing fourteen tons, had been loaded into a boxcar, and reached Tucson safely on the twenty-second of July.[9]

Douglass, visiting friends and relatives in the Boston area, was not expected for two days, so the car remained sealed in the

The Steward Observatory under construction, March 1921.

El Paso and Southwestern yards in Tucson, guarded day and night as it had been during its journey from Cleveland. On Monday, under Douglass's supervision, four large wagons and two heavy trucks moved the twenty cases of telescope parts from the depot to the observatory building on campus. Douglass telegraphed Warner and Swasey that the observatory was now ready, and asked for the firm's H.L. Cook, who reached Tucson on the thirtieth, to superintend the assembly of the instrument.[10]

The housing for the Steward telescope had been ready for several months, awaiting the arrival of the observatory machinery. Located in an isolated portion of the campus, the building presented an arresting picture. The two-story structure was octagonal in shape, with its interior almost completely devoted to the telescope itself. Two small offices and a few other rooms composed the remainder of the floor space. The color of the observatory was also unusual. Although University President von KleinSmid wanted the building constructed of the same red brick as the other buildings on campus, Douglass successfully argued that a white exterior would provide better insulation and keep the telescope in adjustment with less difficulty.[11]

Unpacking and final construction took place on 1 August. Under Cook's direction, the telescope mount was installed on the third, with the tube hoisted into position six days later. Electrical work began on the tenth and was completed and tested by the fifteenth. Minor work continued for the rest of August. Douglass first used the instrument on the twentieth, observing the bright star Sirius during the late morning with the four-inch finder telescope. The only real difficulty was with the clock drive, found to have an error of sufficient magnitude to disrupt accurate photographic work. Cook corrected the malfunction in early September. At this point, the university made the final payment of $8400 to Warner and Swasey. The Steward Observatory now stood complete except for the telescope's optical system, the center of which was the thirty-seven-inch mirror.[12]

On January 4, 1921, Spencer removed its glass disk from the annealing oven, having placed it there the month before. Unfortunately, the disk contained a large crack. The Buffalo firm unsuccessfully tried to weld the disk together in an electric furnace. Spencer cast another disk in late July, after delays caused by problems with the electric furnace. Shortly after placing this new disk in the annealing oven, a severe storm struck the Niagara Falls power plant, burning out a transformer and leaving the furnace without power for some eleven hours. During those few hours the temperature in the oven dropped seventy-five degrees, ruining the mirror blank. In mid-August, the firm cast another large glass, but disaster struck again. Strains from nonuniform temperatures in the electric furnace caused a severe crack in the disk.[13]

The Spencer Company now abandoned the General Electric furnace and rebuilt one of their gas furnaces to be used for the delicate work of annealing the Steward glass. Still another disk was cast. As added insurance, the company also rebuilt an electric furnace to their own specifications and planned to cast a second large disk for that furnace. Of two such disks, one should be perfect. During November and December of 1921, the mirror blanks annealed in the gas and electric furnaces, closely watched by the Spencer staff. The workers turned off the heat in both furnaces a week before Christmas, allowing the disks to cool. On December 22, Douglass received welcomed news from Donald E. Sharp, foreman of the project. Sharp wired: "Telescope disk appears successful in preliminary inspection. Merry Christmas." Although the electric furnace had malfunctioned, causing one disk to crack, the gas furnace performed flawlessly.[14]

Spencer ground and polished the Steward glass during January of 1922, then shipped it to Brashear in Pittsburgh. The preliminary shaping of the mirror took longer than expected. The disk had been cast much oversize, requiring that the diameter be reduced by three inches and the thickness by almost two inches. McDowell finished polishing the Steward mirror in late April and began the final corrections. The mirror gradually approached the required parabolic shape, and by June 6 stood completed. Silvering and adjusting the mirror took another week, after which the smaller mirrors were completed and adjusted. The observatory optics did not arrive in Tucson until July 10, but workmen installed the mirrors in five days. First use of the new instrument came the afternoon of July 17, when Douglass brought the crescent Venus into focus. The Harvard eight-inch telescope was returned the next month.[15]

Over the next few months, Douglass continued his initial observations with the Steward telescope. His earliest investigations concerned the planet Mars during its 1922 opposition. Although no dramatic discoveries came from these observations, they did add to Douglass's large collection of Martian data. Various lunar and planetary phenomena were also studied, putting the new facility to good use. Douglass also plunged into the increasingly demanding field of stellar astronomy, photographing faint nebulae visible from Tucson, conducting photometric work on variable stars, and studying close double stars.[16]

Douglass also spent a great deal of time and effort arranging for the formal dedication of the Steward Observatory planned for April of 1923. Anxious to have a famous astronomer deliver the major address, he turned to George Ellery Hale. Hale had been in Europe for eight months trying to regain his health and would not return until the fall. When neither W.W. Campbell, director of California's Lick

A.E. Douglass inside the Steward Observatory dome, April 1922. The 37-inch mirror was installed four months later.

Observatory, nor Walter S. Adams of the Mt. Wilson Observatory could attend, Douglass arranged for R.G. Aitken, assistant director at Lick, to deliver the major dedication address.[17]

The dedication ceremonies were held at the observatory on April 23 before an audience of several hundred persons. University president Cloyd Heck Marvin opened the program with a welcoming address. He emphasized the appropriateness of an observatory being one of the first major gifts to the university, and concluded with the remark: "Through science . . . our students are going to carry on the traditions of their pioneer forebears in the discovery and definition of the frontiers of investigation." V.M. Slipher, director of the Lowell Observatory, also welcomed the new facility, pointing out the superior climatic conditions found in the region and predicting a great future for the observatory. "Here under the favorable skies of Tucson," he stated, "and with the sympathy and support of this stalwart University and in the hands of its able director, this new Observatory is assured a large and fruitful life."[18]

R.G. Aitken then presented the major address. He began with the observation that the dedication of the Steward Observatory was taking place during the 450th anniversary of the birth of Nicholas Copernicus, in whose ideas the geocentric universe met its most famous challenge. He briefly surveyed some of the more significant chapters in the history of astronomy, and then discussed in detail the determination of stellar distances and recent research concerning stellar evolution. Concluding his comments, Aitken stressed the important role the Steward Observatory would play in future developments in astronomy, providing both research and education as a scientific outpost in the southern Arizona desert.[19]

Douglass delivered the final address of the evening, recounting the trials and difficulties in obtaining the observatory and its equipment. At the end of his remarks he took a few moments to consider the future of the facility. He emphasized the importance of research to the growth and reputation of the observatory. "I want this Steward Observatory to live," explained Douglass, "and in living it must grow, and in growing it must produce results." There was only one way for this to happen: "Its use for classes is fine; its use for the public is fine; but it will not live without scientific results." Douglass stressed the need for new equipment and scientists to make optimal use of the observatory, and sought to convince his audience of the value of such investments:

> Scientific research is business foresight on a large scale. It is knowledge obtained before it is needed. Knowledge is power, but we cannot tell which fact in the domain of knowledge is the one which is going to give the power, and we therefore develop the idea of knowledge for its own sake, confident that some one fact or training will pay for all the effort. This I believe is the essence of education wherever such education is not strictly vocational....In this Observatory I sincerely hope and expect that the boundaries of human knowledge will be advanced along astronomical lines....

> Historically and practically, therefore, the work of this Observatory is a line of research worthy of a rapidly growing State and University, of our fine climate, and of our aggressive and intelligent people.[20]

With the facility formally dedicated, Douglass turned his attention to acquiring additional equipment. He particularly wanted a spectroscope and a student telescope, each estimated to cost $5000 or more, but had no success in obtaining either instrument. Acting on another of Douglass's requests, the university hired a second full-time instructor for the astronomy department in 1925. The new employee was Edwin F. Carpenter, who had bachelor's and master's degrees from Harvard (where he had been elected to Phi Beta Kappa), and a doctorate from the University of California. Carpenter took charge of most of the class work, leaving Douglass free to pursue his varied research interests.[21]

The Solar Eclipse Expedition of 1923

The Steward Observatory's earliest major research project involved an expedition into Mexico to observe the solar eclipse of September 10, 1923. Douglass had been planning for this eclipse for five years, obtaining information concerning the eclipse path and collecting appropriate instruments. Financing for the expedition came from a $500 budget arranged by President Marvin the previous year.[22]

Careful scrutiny of weather patterns and accessibility convinced Douglass that the best location for observing the eclipse would be Puerto Libertad, on the east coast of the Gulf of California. This location would be relatively inexpensive to reach and lay only eighteen miles north of the central line of totality.[23] Douglass next collected instruments for the expedition, borrowing a five-inch lens from the U.S. Naval Observatory in Washington, D.C., and several mirrors from the Carnegie Desert Laboratory in Tucson. He also secured the services of Godfrey Sykes, the mechanic at the lab, who contributed an eight-inch lens with a focal length of more than fifty feet. The various pieces of optical equipment were combined to form astronomical cameras to record the progress of the eclipse.

With the needed equipment in hand, Douglass arranged for transportaion to the Puerto Libertad site. On Saturday, September 1, Douglass sent the expedition's large truck toward Mexico. The next morning, Douglass and his assistants drove south in a seven car caravan and crossed the international border at Sasabe in mid-afternoon. Later, during a heavy rain storm, the party became lost and bogged in deep mud, but local residents helped them on their way. Catching up with the truck beyond San Rafael, the expedition slowly made its way to the gulf site, reaching Puerto Libertad at noon on September 6.[24]

As the expedition unloaded its equipment, Douglass discovered that the Naval Observatory lens had been left on his desk at the university. Glenton Sykes (Godfrey's son) and William Done, an astronomy student, left immediately for Tucson in a stripped-down Ford. Reaching Altar the next day, they telegraphed President Marvin to send the needed lens toward them in a university car. Sykes and Done reached the border five hours later, just as the university car approached from the north. They returned to camp with the lens in the afternoon of September 8.[25]

Douglass and his friends erected their equipment on September 7 and 8. The hot and humid climate attracted droves of flies and mosquitoes which made the slightest task almost intolerable. On Sunday afternoon an extended thunderstorm forced Douglass to dismantle some of the instruments, destroying the careful adjustments of the previous two days. Drinking water from a nearby well also posed a problem. Members of the expedition removed "some small dead animals" from the well and bailed it out, but the water remained no more than potable. A few days later, they discovered a fresh spring nearby at low tide.[26]

As the hour of the eclipse appproached, the expedition made final preparations. Douglass relied on the published maps of the eclipse path, instead of rigorously calculating the exact time of eclipse at the Puerto Libertad site. Because of this, the eclipse began ninety seconds earlier than expected. Photographing the eclipse was tedious. Douglass had to work in a darkened protective covering to handle the photographic plates. Because he was in total darkness, Douglass had to rely on word from outside observers to determine the time at which to start making photographic exposures. The other expedition members, however, had never observed a solar eclipse, and became so fascinated by the phenomenon that ten to twenty seconds elapsed before anyone told Douglass that the eclipse had begun. The combination of these problems and the many tourists from nearby areas limited the number of exposures taken. The expedition started back to Tucson on September 13 and arrived late the following day.[27] Douglass and his crew spent two days developing the photographic plates, after which Douglass went to California to attend scientific meetings. His report and photographs were well received there, with an expanded acount published the next year in the *Publications* of the Astronomical Society of the Pacific.[28]

A few months after his eclipse expedition, Douglass turned once again to the planet Mars. Determination of the true color of the markings on Mars could be a valuable addition to the investigation of possible life on the planet's surface. Douglass's notes on his 1924 observations stressed the appearance of the surface features. The green color of these markings was "very evident" and "unmistakable" throughout the observations of late summer and fall. The use of twenty-three infrared photographic plates resulted in improved views of the planet's surface and confirmed the presence of the dark markings visible in normal light. His photographic work also displayed evidence of clouds, leading Douglass to conclude again that the Martian atmosphere was in many ways similar to that of Earth, an interpretation later proved erroneous. Observations of the Martian oppositions during the rest of the decade reinforced Douglass's view that the planet's dark surface markings represented a vegetative life form, a belief shared by most astronomers of the period.[29]

Planetary research continued as the major topic at the Steward Observatory during the 1930s. Although most of the world's observatories emphasized stellar and galactic work, Douglass and his assistants studied such diverse topics as the lunar surface and the recently discovered planet Pluto. They recorded the Martian oppositions of 1931–32, 1933, 1935, 1937 and 1939, but the pressures of other research (Douglass wanted the Steward program to cover as wide a variety of astronomical topics as possible) made a systematic investigation of the red planet impossible. At the end of the decade, though, the observatory hosted a visiting astronomer who used the facility for detailed Martian study. E.P. Martz, who had extensive experience at the Mount Wilson and Griffith observatories, secured a temporary position at the university to continue his heterochromatic photography of Mars. During the 1939 oppositions, Martz photographed the planet using various filters to show Mars in different colors. Coupled with his photometric measurements of color intensities, his photographic work assisted in the mapping and analysis of Martian surface features.[30]

Douglass and his staff also pursued stellar and galactic research. In 1925 V.M. Slipher, director of the Lowell Observatory, asked Douglass if his facility would participate in an international program of nebular photography. For the past dozen years, Slipher had been studying the spiral nebulae, finding that they possessed very large radial velocities. This discovery suggested that there were bodies external to the Milky Way galaxy. During the mid-1920s, Edwin P. Hubble of the Mt. Wilson Observatory searched the nebulae for stars known as Cepheid variables, whose luminosity and periodic fluctuations could be used to determine their distance from the earth. By 1924, Hubble had calculated the distance of several nebulae containing Cepheids, concluding that the nebulae were phenomena that existed at great distances from the Milky Way galaxy and, in fact, were galaxies themselves.[31]

Although Douglass was eager to undertake work of this kind, he had no funds to hire an additional assistant, which delayed Steward participation in this significant program for three years. Finally, in 1928, E.F. Carpenter agreed to add this extra assignment to his already active schedule and directed the photographic program. Focusing on the sky just south of Tucson, Carpenter regularly made exposures until the end of the project in 1932. An attempt to help determine the actual dimensions of the Milky Way, Carpenter's research led him to theorize that the galaxy was actually much smaller than most astronomers thought. He believed that clouds of light-absorbing material in space made the extragalactic nebulae appear farther away than they actually were, an illusion which he thought caused his colleagues to overestimate the dimensions of both the galaxies and the universe. Carpenter's analysis later proved inaccurate.[32]

By the early 1930s, Douglass's dedication to the establishment of astronomy in southern Arizona had produced encouraging results. The Steward Observatory, beginning its second decade of operation, had entered the increasingly exciting and productive world of American astronomy, a field of growing primacy to the world's scientific community. Yet Douglass had not spent all his years at the University of Arizona peering through the telescope and planning future projects. He was, after all, an active member of the faculty with many teaching and administrative duties, and as an increasingly visible member of an expanding university in a growing community, he found himself very much a part of Tucson and university society. From the very beginning, Douglass's outgoing personality and commitment to education had combined to keep him in the public spotlight.

6

A Public Figure

AFTER HIS DISMISSAL from the Lowell Observatory in 1901, Douglass attempted to maintain his academic and scientific position in Flagstaff. He realized from the beginning, however, that a connection with an established institution was essential to his professional status. For several years Douglass searched for employment without success. Finally, near the beginning of Douglass's eleventh Arizona winter, his fortunes began to change.

In late December of 1905, Douglass wrote to Kendrick C. Babcock, president of the University of Arizona in Tucson, seeking a teaching position. Fearing that the proposed joint statehood of New Mexico and Arizona would lead to the abandonment of Northern Arizona Normal, he described his educational and teaching background, and applied for whatever appointment Babcock could offer. A few days later, Babcock replied that he needed a physics instructor, but that any successful applicant would also have to teach a course in physical geography. Douglass quickly responded with details of his work in physics and astronomy, and indicated that he would be happy to teach physical geography as well. By late January of 1906, Douglass had accepted Babcock's informal offer of the physics post, and added that the possible future chairmanship of the department mentioned by Babcock would present suitable compensation for the financial loss to be suffered by leaving Flagstaff.[1] By May, Douglass received formal appointment as Assistant Professor of Physics and Geography, at an annual salary of $1200. His teaching duties would include sophomore physics (including laboratory work), beginning Spanish, physical geography, and preparatory physics, the last course geared toward the high school students who composed a significant portion of the enrollment.[2]

Douglass anticipated several disadvantages in joining the faculty at the University of Arizona. In a long letter to Babcock, he emphasized that his move involved a significant financial sacrifice. In the first place, he would be accepting a lower salary than the $1400 offered by the Normal School. Of equal importance, Douglass told Babcock that in accepting the Tucson appointment, he had ignored the assurances of several board members that he would be the next president of the Flagstaff institution. Nonetheless, Douglass welcomed the opportunity to teach at an established university, despite its small enrollment of 215 students, a twenty-six member faculty, and a library of fewer than 10,000 volumes.[3]

Douglass recognized that his academic career would require continued diligence and the reestablishment of his professional credentials. One major shortcoming was the lack of an advanced degree, a problem which he took steps to correct while still in Flagstaff. Shortly after his appointment to the University of Arizona, Douglass wrote to F.S. Luther, who had recently become president of Trinity College. Considering his past work, especially with the Harvard and Lowell observatories, Douglass hoped that the Trinity Board of Directors could be persuaded to give him an honorary advanced degree. This would in some respects compensate for his lack of graduate training, a valuable commodity in the academic world of 1906. Luther replied that Trinity's board had closed all such actions at their April meeting, but assured his former student that the college would gladly grant him an M.A. *honoris causa* at some future date.[4]

After moving to Tucson and starting classes in the fall of 1906, Douglass continued his pursuit of an advanced degree. In late October he wrote President C.W. Eliot of Harvard and asked to be considered for a Ph.D. the following June. He had talked to the president of the University of California, Berkeley, who assured Douglass that with work at Lick Observatory similar to Douglass's at the Harvard Observatory, the Berkeley institution would grant the doctorate. Early the next month, the secretary of Harvard's graduate school responded that Harvard could not grant Douglass a Ph.D. because his observatory work had not been performed under a faculty department conferring doctorates.[5] During the 1907–1908 academic year, however, Douglass finally succeeded in his campaign. F.S. Luther arranged for the Trinity trustees to grant Douglass the honorary degree of Doctor of Science. Luther presented the degree at commencement in Hartford on June 24, 1908. The Arizona astronomer would be called "Doctor Douglass" for the rest of his long life.[6]

In the fall of 1906, Douglass began teaching classes in college and preparatory physics, Spanish, and physical geography, with a total enrollment of sixty students. He continued the physics and geography courses the following spring, but substituted a course in physical measurements for Spanish. This offering included half a semester's work in hydraulics, a valuable course for the many mining and engineering students enrolled on campus.[7]

The Board of Regents showed its satisfaction with Douglass by promoting him in May of 1907 to Professor of Physics and Astronomy and Head of the Department, at an increased salary of $1800 for the year beginning September 1. The inclusion of "Astronomy" in his title had been an early suggestion of Douglass, made in the hope of gaining support for an observatory in southern Arizona. For the next four years, Douglass's class work remained fairly constant, consisting of one or two astronomy classes and as many as three physics classes per semester. Total enrollment for his astronomy and physics courses ranged from a high of forty-four in the spring of 1908 to a low of twenty-six the following fall. Despite the large variance from year to year, semester enrollments seldom changed to any great extent, indicating Douglass's popularity with students.[8]

Despite his secure and improving position at the University of Arizona, Douglass continued to search for more professionally advantageous employment which would return him to active work in astronomy. During his first Christmas vacation in Tucson, he wrote to President John M. Baer of Occidental College in Los Angeles, applying for the directorship of the college's new observatory. Informed that Occidental's small facility did not require a full-time director, Douglass immediately applied to W.W. Campbell at Lick Observatory for any post that might be available, but Campbell could offer his friend no hope for a position. A.N. Taylor, president of Northern Arizona Normal, offered Douglass the chairmanship of the science department in February, but the latter remained convinced that his position in Tucson would prove far more beneficial than a return to Flagstaff.[9]

The next academic year found Douglass involved in further attempts to leave Tucson. In November, he asked President G.F. Bovard at the University of Southern California for details concerning a proposed chair in astronomy. Bovard answered Douglass's application with the disheartening news that no action could be taken until February of the next year, but if any suitable opening occurred he would write him. Douglass later applied to George Ellery Hale at the Mt. Wilson Observatory, east of Los Angeles, hoping to gain a position with the most noteworthy astronomical facility in the Southwest. Hale could give Douglass no encouragement either. Although the sixty-inch telescope, the largest in the world, was near completion, the observatory itself had no funds to expand its staff.[10]

For all his fourteen years in Arizona, Douglass remained a New Englander. In a letter to his mother written December 13, 1908, he expressed a profound homesickness for his native region. He and Ida had been considering buying a house in Tucson, but he confessed to some ambivalence on the issue. "I would not like to do anything," he wrote, "to tie me out here in the face of any opportunity to come East to live." Douglass confessed that "I don't see any such chance just at present," but nonetheless very much wished to return. "I can see how the Pilgrim Fathers felt," he concluded, "when they fairly committed themselves to the New World...." His desire to move back east, however, remained an unfulfilled wish. His only offer of a transfer came from Flagstaff friends who wanted to put Douglass's name forward for the presidency of Northern Arizona Normal in March 1909.[11]

In the summer of 1910 the Douglasses enjoyed a temporary change of scenery when they journeyed to Europe. From late June until until early August, Andrew visited various observatories and laboratories, while Ida studied French at the Sorbonne. Visiting Paris in late July, Douglass inspected the Observatory of Paris and the Sorbonne physics laboratory, and had tea at the home of the great French astronomer Camille Flammarion. Douglass left Ida to finish classes while he traveled to London and inspected the Solar Physics Observatory at South Kensington. He also visited the Royal College of Science (established in 1881 as a government-supported teaching institution) as the guest of the eminent British astronomer Sir Norman Lockyer and his son, meteorologist W.J.S. Lockyer. Returning to the United States, Douglass stayed several weeks at the family home in East Wareham, Massachusetts, and returned to Tucson for the beginning of classes in September. Ida returned from Europe a few weeks later.[12]

Teacher and Administrator

Douglass's fifth year at the University of Arizona promised no unusual developments. Teaching and research occupied his time as they had in earlier years. As the fall progressed, however, the campus community began to change. On October 22, 1910, President Babcock publicly announced his resignation from the university. He had recently been appointed to a specialist's post in the Education Bureau of the Department of the Interior, a position that would allow him to conduct his own research. Babcock planned to leave shortly before Christmas.[13]

The Board of Regents met on Friday, December 20, concluded that a careful search for a new president would take time, and decided to choose an acting president. They had little difficulty in selecting a candidate, for Douglass had acted in Babcock's stead during the latter's frequent absences.[14]

Douglass's tenure as university president proved relatively quiet, allowing him to continue his teaching duties. University enrollment stood at 195 at this time, with thirty faculty members, seventeen buildings, and a library housing almost 16,000 volumes.[15] During the next five months, Douglass oversaw the operation of the university while the Board of Regents screened applicants carefully. After considerable investigation, the regents invited Arthur H. Wilde to be president of the University of Arizona. Resigning his post as assistant to the president of Northwestern University, Wilde arrived in early May to assume his duties. He was inaugurated on the evening of May 17, at which time Governor Richard A. Sloan gave the major address. Wilde's first official duty occurred on June 1, when he presided over the 1911 commencement exercises. Retiring interim President Douglass gave the commencement address.[16]

Stepping down from the presidency at the beginning of the summer of 1911, Douglass spent the next few months on long-neglected research projects. That fall, he returned to teaching physics and astronomy courses, which enrolled from thirty-five to forty students each semester. Douglass took great care in his teaching, preparing comprehensive lecture notes and supervising the class experiments in physics. To make his introductory astronomy courses even more interesting, he included significant insights into the history of astronomy, showing his students the development of the knowledge of the universe and providing them with contemporary facts and theories. Requiring a moderate degree of mathematical sophistication in his more advanced classes, Douglass gave university students a valuable background in astronomy and physics, equal to that he had enjoyed as an undergraduate.[17]

In December of 1911, Douglass applied to President Wilde for a sabbatical year. Although he had been on the university faculty only six years (instead of the usual seven), Douglass wanted time to continue his research by affiliating himself temporarily with a major German or American observatory. He also wished to obtain information on the methods used by large American universities in physics classes. Douglass wrote to several colleges in southern California to apply for part-time teaching positions for a portion of his absence. Neither the University of Southern California, Occidental College, nor

Throop Institute (later California Institute of Technology) could offer him a position.[18]

Emphasizing the exceptional nature of their decision, the regents approved Douglass's sabbatical request in mid-January of 1912. Requiring a pledge that he would return for the next school year, they granted the customary three-fifths salary, citing his loyal service to the university. Unable to find an expedition traveling to Argentina to observe the October solar eclipse, Douglass attempted to obtain a position with the Yerkes Observatory. The director, Edwin B. Frost, told Douglass he would be welcome at the facility, but could offer no remuneration for his services. By mid-September, Douglass had decided to spend a few months visiting fellow researchers and facilities in Germany, while Ida undertook graduate work at Radcliffe. He planned to return in early 1913 and work at Harvard and the Carnegie Institution for the rest of his leave, returning to Tucson in the fall.[19]

Douglass sailed from New York and reached Liverpool on October 9. He spent the next three weeks in England visiting observatories and museums. Douglass next traveled to Germany, arriving in Berlin on October 29, and confiding to his mother the next day that he greatly enjoyed Germany and its "go," which reminded him of the United States. Admitting that his knowledge of German remained inadequate, he indicated his success in conversing with the natives by confessing, "Usually they send for someone who can speak English."[20]

Douglass remained in Germany until a few days before Christmas. He traveled widely, visiting museums and observatories and collecting tree samples to complement those he had gathered earlier in England. Visiting a number of German schools and universities, he attended a class lecture by the famous scientist Hugo von Seeliger, one of the first astronomers to suggest an elongated shape for the Milky Way. Douglass's German visit provided great intellectual excitement. Writing a college classmate several months later, he described the German museums as "regular hot houses of professional industry where your ideas grow in spite of anything."[21]

Leaving Berlin for Copenhagen on December 22, Douglass remained in Scandinavia until early January, collecting research material from Sweden and Norway, and then went on to

London to attend a meeting of the Royal Astronomical Society on the tenth. He sailed a few days later for the United States. On reaching New York City, he hurried north to his mother's home, arriving in Newton Highlands, Massachusetts, only a few weeks before her death. Although in mourning, Douglass spent the next seven months organizing his research material and preparing for the fall semester at the University of Arizona.[22]

For the next quarter century, Douglass taught astronomy at the university and, until the end of World War I, also conducted the physics classes. During these years, he sought to stimulate his students to become more curious about the world and universe around them. He used telescopes and other instruments whenever possible and conducted laboratory experiments designed to duplicate aspects of the great scientific achievements of the past. Teaching a university summer school course in Flagstaff in 1920, Douglass arranged for the use of the Lowell Observatory to increase his students' knowledge of astronomical instrumentation. Three years later, he drafted plans for a motion picture (still somewhat of a novelty for academic purposes) to illustrate work in astronomy, discussing instruments as well as celestial bodies. The many demands on his time, however, prevented Douglass from completing this project.[23]

In the fall of 1915 Douglass again became actively involved in the administration of the university, when he became Dean of the new College of Letters, Arts, and Sciences. The university enrollment had reached 300 and Douglass's appointment to the new post symbolized the high regard in which he was held by the faculty and administration. That the student body shared this esteem had been evidenced a few months earlier by the dedication of the university yearbook to Douglass.[24]

Douglass's tenure as dean proved relatively uneventful. As routine administrative matters took little of his time, he continued his teaching and research. The only burdensome period occurred during January of 1916, when Douglass assumed the formal functions of President von KleinSmid while the latter traveled in the eastern United States.[25] As time went on, Douglass's research interests took increasing amounts of his time, so in February of 1918 he asked von KleinSmid to relieve him of the dean's post. The regents accepted his resignation, thanking him for his laudable service and continuing him as

Professor of Physics and Astronomy. They also granted Douglass's request that he be appointed director of the new Steward Observatory.[26]

Despite his influential position at the University of Arizona during these years, Douglass continued to seek opportunities elsewhere. While serving as president, he applied to the University of Virginia for a position in their astronomy department. Two years later he wrote to his friend Kendrick C. Babcock, soon to be a dean at the University of Illinois, asking for assistance in securing a position in the astronomy or physics departments at the midwestern school. By late 1913, Douglass was also writing colleagues and friends on the east coast inquiring about possible openings. In all his applications, Douglass stressed his wife's inability to cope with the Tucson weather, especially during the summer. Writing to Trinity classmate Boardman Wright in June of 1914, Douglass asked if any New York City institutions had openings. He confided "It is true that I have splendid scientific opportunities out here but I want to get into a bearable climate."[27]

Douglass's desire to leave the university took a more serious turn during the summer of 1915. Word came that a new women's college would soon open in New London, Connecticut. Douglass telegraphed F.H. Sykes, the college president, asking to be considered for the professorship of mathematics and physics. Sykes asked for full particulars, including expected salary, which Douglass enthusiastically forwarded. Douglass also wrote to F.S. Luther for advice. Luther viewed the Connecticut College for Women as an experiment which showed promise for ultimate success. In the immediate future, however, the college would probably have difficulties which would exert a hardship on faculty members. "If you were thirty instead of a good deal more than that," wrote Luther, "I should feel that it was a chance worth taking. As it is, unless you are considerably dissatisfied where you are, I should remain." Luther's advice proved prophetic, for Sykes could only offer Douglass a half-time position.[28]

Organizing a Scientific Union

Although the local climate remained the major reason for Douglass's attempts to leave the University of Arizona, the area's isolation from scientific organizations was equally disagreeable. After the end of World War I, however, Douglass and other scientists in the American Southwest began discussing the organization of a Southwestern Division of the American Association for the Advancement of Science. They conducted inquiries to determine the interest for such an organization, and on April 10, 1920, held an organizational meeting in Tucson on the University of Arizona campus. The meeting included President von KleinSmid, Dr. Elliott C. Prentiss (a physician from El Paso who had first proposed the organization), and Dr. D.T. MacDougal of the Carnegie Desert Laboratory in Tucson, who also served as the delegate from the National Executive Committee of the American Association. Six other scientists from Arizona, New Mexico and west Texas completed the meeting's membership.

The group elected Prentiss and Douglass chairman and secretary, respectively. An election was then held for officers. The eminent archaeologist Edgar L. Hewett, of the School of American Archaeology in Santa Fe, became the first president. Prentiss would serve as vice president and Douglass as secretary-treasurer.[29]

The new division grew steadily over the next several months. By late October, it had attracted the region's scientists, even securing the services of Vesto Melvin Slipher, director of the Lowell Observatory, to serve as chairman of the physical sciences section. At a meeting of the Executive Committee the day before the group's first convention, Douglass reported a membership of 143. The El Paso conference, which began December 2, proved very successful, and was capped by the election of A.E. Douglass as president of the Southwestern Division for 1921.[30]

During Douglass's presidency, the division served as an important means of communication among the scientists of the Southwest. In addition, it mounted a campaign to protect the Picture Rocks northwest of Tucson as a national monument. These rocks were covered with petroglyphs made by the ancient Indians of the area, and presented one of the few examples of this art in the vicinity. As president, Douglass wrote to the director of the National Park Service, asking him to investigate the matter. Although unsuccessful in obtaining national monument status for the rocks, the campaign nonetheless showed that the Southwestern Division had become a viable organization.[31]

By mid-1921, the executive committee of the division had selected Tucson as the site for the second annual meeting. The meeting was held during the second half of January on the University of Arizona campus, whose hospitality R.B. von KleinSmid had generously offered. As division president and the university's most distinguished scientist, Douglass became the focus of the meeting, arranging tours of the science facilities on campus and displaying his own research. He also presented papers to the Physical and Social Sciences sections and, on Thursday evening, January 26, delivered his presidential address entitled "Some Aspects of the Use of Annual Rings of Trees in Climatic Study." The address was later published in *The Scientific Monthly* and the *Smithsonian Report* for 1922.[32]

Douglass remained active in the Southwestern Division for the next thirty years, serving on committees, presenting papers and chairing sessions. The papers he delivered reflected the catholicity of his scientific research and were always well received. In the February 1926 meeting in Phoenix, for example, Douglass presented three papers. He spoke on "The Infra-Red Photographs of Mars and its Atmosphere," the "Interpretation of Unwritten History," and "Terrestrial Response to Solar Radiation."[33]

Douglass's position as director of the Steward Observatory led him to further involvement with the public. Douglass early made the observatory available to the community. While the observatory was under construction, he gave a series of open-air lectures on such topics as constel- lations, planets, and stellar evolution at the observatory building on Tuesday nights. Public lectures and open evenings continued throughout Douglass's tenure as director of the observatory, often attracting more than 100 visitors. During a viewing session devoted to Mars in August of 1924, 400 interested spectators peered through the Steward telescope at the red planet.[34]

Public Controversies

Relations with the community were not always so peaceful. On January 8, 1924, Douglass spoke before the Women's Service League of Grace Episcopal Church, where he was a member. His address focused on the supposed dichotomy between science and religion as evidenced by the growth of fundamentalism in the United States. This movement advocated a literal interpretation of the Bible and a rejection of all science seen as conflicting with that interpretation. Arguing that the two were not incompatible, Douglass urged his audience to encourage young people to study science and to clear away some of the more mythical elements in religion, such as the belief that the world was created in six days. "We must not ask our young people," Douglass continued, "to put faith in things that are obviously to their clear and educated minds untrue." He proposed a modernization of religion for the present generation, "not accepting the mythical interpretations given, but giving them instead a common sense religion which will help them and upon which they can lean."[35]

While only one of several such discussions in Tucson churches in early 1924, Douglass's appearance before the Episcopal women triggered a heated response. R.S. Beal, pastor of the First Baptist Church in Tucson, had been conducting a Sunday afternoon class in fundamentalism for several weeks. Douglass's address, as reported in the Tucson press, prompted Beal to write an open letter to university President Cloyd H. Marvin, which appeared in the January 11 *Citizen*. Beal did not quarrel with the astronomer's right to believe as he saw fit, but the minister objected strenuously to teaching such a vile doctrine in the classroom. He emphasized his fondness and respect for the University of Arizona, but felt "grieved beyond expression with the bold materialism and gross infidelity of many of its professors...." Beal urged professors like Douglass to examine the other side of controversial questions such as evolution in order to "save from ship wreck on the rocks of infidelity the faith of many of the students." He fired another blast at Douglass on the morning of January 20 in a sermon entitled "The Voice of God on the Millenium, an answer to Dean Douglass."[36]

Debate continued when Douglass wrote two articles for the *Citizen* on "Evolution as Seen by the Astronomer." Again trying to dispel the supposed hostility between science and religion, he discussed the research leading to the currently-accepted age of the universe, much older than the 6000-year figure popular with fundamentalists. Douglass went on to say, however, that the great age of the universe led to "the old, old discovery; God and his heaven are in you and me and in all human hearts, and always have been there." Rather than presenting a harsh attack on religion, modern science showed that creation was a "continuous process lasting throughout the ages and we now call it evolution, believing that it will give opportunity for a new and stronger faith." Evolution, he continued, led to the idea of sacrifice for the future good of the group, "and that is the essence of every religion." Evolution served a valuable function, supplying "in [times] of peace an intellectual basis for religious activity which in times of stress and danger and rapid experience comes through the heart and emotions."[37]

Beal and his supporters evidently believed that arguing specifics represented the best method to refute Douglass and his evolutionary fellow travelers. Several weeks after Douglass's articles appeared, an anti-evolutionist attacked him on "scientific" grounds in the Tucson *Citizen*. One of the examples of Biblical mythology frequently cited by Douglass was the account of Joshua commanding the sun to stand still. Douglass emphasized that the only way this could be accomplished was for the earth to stop rotating on its axis, a phenomenon that would disrupt so many geological and physical forces as to cause certain cataclysmic destruction. Douglass's challenger maintained that even though the earth's rotational velocity approached 88,000 feet per minute, this represented a scant 0.00068 of the planet's circumference. The writer, displaying a grossly inadequate knowledge of geophysics, concluded that the centrifugal force resulting from the earth's rotation would be practically nil and stopping the earth would have little effect. Douglass did not dignify this idea with a reply. Tucson's evolution debate lost momentum during the mid-1920s, although Reverend Beal attempted in 1927 to secure an anti-evolution law from the Arizona legislature without success.[38]

Difficulties with the community did not always concern scientific theories. Despite the isolated location of the Steward Observatory site in 1916, Tucson's postwar construction (which Douglass had not anticipated in his selection of a campus location for the telescope) quickly presented problems. In 1920, Douglass told President von KleinSmid that off-campus building activity north and east of the observatory was causing an increase in light which, if it continued unchecked, would seriously hamper the observatory's photographic work. He informed the president that he would take the matter to the city council. Douglass also recommended that the university plant foliage around the observatory to block out ground lighting and hold down airborne dust. The city council cooperated with the observatory, instructing the city engineer to confer with Douglass before erecting city lights near the observatory. Lights that had to be installed were carefully shielded. By the end of the decade, Tucson's growth had greatly increased the lighting problem. In his 1928 report, Douglass emphasized that the increase in city lighting was hurting long-exposure photographic work and would eventually require the removal of the Steward reflector away from light interference. Thereafter he kept close watch over development in the observatory area.[39]

The 1930s witnessed a virulent debate between the observatory and the business community. When the city council received petitions seeking permission to install ornamental lighting in the area southeast of the observatory, the city engineer consulted with Douglass to minimize interference with Steward research. Of greater potential danger was the plan to rezone the area directly south of the installation for business. In a written plea to the mayor and council, Douglass warned that bright store fronts would seriously jeopardize the photographic work of the observatory. He pointed out that a street lamp three blocks from the telescope presented an image three million times brighter than many of the objects photographed by the Steward instrument. The editors of the two Tucson newspapers joined Douglass in urging the city and citizens to do everything possible to minimize the interferences with Douglass's scientific

A.E. Douglass at the Steward Observatory, 1941.

work. These efforts met with success during mid-February of 1931, when the council rejected petitions for increased lighting and zoning changes near the Steward Observatory.[40]

Efforts to rezone a street corner a few blocks northeast of the observatory resulted in a heated debate the following year. At the city council meeting of June 20, 1932, a group of businessmen proposed rezoning the corner of Speedway and Campbell Avenue for business. Douglass made his usual argument against such proposals, again stressing the need to minimize light interference. The council voted to deny the rezoning request. Two years later, the dispute resurfaced when a new attempt to gain business classification was begun by the same group. Douglass's arguments again carried the day, but rezoning attempts continued.[41]

On April 13, 1936, the Tucson City Council considered yet another request to rezone the intersection. Throughout the week preceding this meeting, Douglass, Carpenter and President H.L. Shantz met to determine the best course of action. Even though the attorney for the businessmen who requested the rezoning asked the university to drop its interest in the matter, Shantz urged Douglass and Carpenter to explain to the council the possible damage to research from increased lighting. At the council meeting, a petition signed by thirty-one members of the neighborhood who opposed the rezoning was presented. None of the signers had any connection with the university. Both the Board of Adjustment and the City Planning Commission recommended that the petition for rezoning be denied.

This latest success for Douglass proved particularly vexing for one of the petitioners, James R. Dunseath. Three days after the council meeting, Dunseath wrote Douglass, accusing him of obstructing the good of the community for reasons of "self-aggrandizement." Continuing his attack, Dunseath asserted:

> You have caused severe losses to be sustained by myself and others on several occasions and have blocked the growth of the City of Tucson, and I am now of the opinion that a little advertising with the Board of Regents...will open their eyes to what a few, at least, of the taxpayers

> who are contributing to your salary and have thereby made your investigations possible, think your activities will mean to the University later.

Dunseath's threats failed to materialize, but such hostility reinforced Douglass's belief that the only adequate solution was to move the Steward Observatory to a more secluded spot.[42]

The University of Arizona itself often showed a disregard for astronomy. Beginning in 1920, Douglass tried to reduce the amount of dust in the observatory area. In addition to improving atmospheric quality, the lessening of dust would reduce the amount of resilvering necessary to keep the main mirror in optimum working order. Traffic, stables, cavalry practice, and athletic fields all remained in the area, however, causing the astronomical work of the observatory to suffer. Expansion of university facilities also threatened the Steward Observatory. Without consultation with the astronomy staff, the physical education department laid out a five-hole golf course around the observatory in early 1929. One of the holes was adjacent to the four-inch telescope used for solar work and student observations. After several cases of near damage, the observatory dismantled the telescope and discontinued its use as a teaching tool. The later construction of a small building to house the instrument eliminated the immediate difficulty, but Douglass urged President Shantz to include a member of the observatory staff on the university planning committee, so that the astronomers would know beforehand of any proposed changes in the observatory area. This suggestion brought no response.[43]

Douglass and his staff continued their fight to maintain suitable conditions for astronomy. The university developed plans for a new infirmary in 1935 and 1936, again without knowledge of the observatory staff. The proposed building would be erected 150 feet north of the telescope, where discharges from the infirmary's heating system would seriously hinder astronomical observations. Douglass managed to convince the planners to locate the medical facility on a site less hazardous to astronomy. No sooner had this been accomplished, however, than the Steward director learned that officials planned a new auditorium for the area near the observatory.[44]

To combat this latest threat, Douglass relied on his professional association with the preeminent astronomers of the nation. He wrote to Harlow Shapley of Harvard, as well as to the directors of the Mt. Wilson, Yerkes and Lick observatories, requesting their opinions of the proposed auditorium. Without exception, these astronomers echoed Douglass's sentiments that a new building would jeopardize the work of the Steward Observatory. W.H. Wright of the Lick Observatory told his Tucson colleague: "It is difficult for me to imagine that the University would consider a measure, the effect of which would be to imperil or destroy such a scientific asset." Shapley stated that the location of an auditorium near Steward "would be essentially ruinous for photographic research." After receiving these letters, Douglass wrote to the administrators of the university, emphasizing that the observatory's work was very important to science and represented far more than "idle star gazing."[45] His efforts once again brought success, and the auditorium was built at the other end of campus.

University students reflected even less appreciation for the Steward Observatory than did the administration and the community. During the spring of 1929, enterprising students stole the canvas cover of the four-inch telescope located on the observatory grounds. Although woefully inadequate, this cover had been the only protection for the instrument. The golf course continued to cause difficulties as well. In early May of 1930, Douglass was loading his automobile for a lecture when a golf ball hit the wheel of his vehicle. The young woman whose drive had gone astray said that it was fortunate that the ball did not hit a window, to which Douglass replied that she would have had to pay for the damage. The golfing coed answered: "[N]ot at all; our instructor has told us we have the right of way here and if damage is done we are not responsible. . . ." Douglass's suggestion that the observatory antedated the golf course made no impression on the golfer. His warning to university administrators that a golf ball striking the main mirror would cause $10,000 damage failed to bring about any changes. Vandalism was also a problem. Several years later, a fraternity group returning from a basketball game thought it would be amusing to throw a small rock into the telescope while astronomers were making observations. Fortunately, the projectile lodged on a ledge inside the tube, preventing the rock from striking the mirror.[46]

Douglass increasingly believed that the Steward telescope would eventually have to be moved away from the city. In 1930 he found a location twelve miles east of town near the proposed Cactus Park, which soon became the Saguaro National Monument. This isolated area had low hills which would protect the telescope from atmospheric currents and city lights. The Steward director developed two plans, the most likely involving the move of the thirty-six-inch instrument to the Cactus Park site and its replacement on campus with a ten- to twelve-inch refractor. This proposal would cost some $50,000. The more ambitious plan would leave the Steward telescope on campus and construct a new sixty-inch instrument at Cactus Park at a total cost of over $300,000. The university gained title to the Cactus Park site, but financing the move proved impossible.[47]

Despite difficulties with the university and community, Douglass's first three decades in Tucson had proved incredibly rewarding. His services as a teacher, university administrator and scientist complemented each other and made him a prominent figure in southern Arizona. Symbolic of his widespread activities was Douglass's work to establish and maintain the Steward Observatory. With that facility, the University of Arizona joined the select group of colleges with programs of serious astronomical research, giving it the publicity Douglass had long predicted. At the time of his retirement as director in 1937, Douglass could look back on many years of the encouraging growth of southern Arizona astronomy. In typical Douglass fashion, however, he was once again looking ahead, this time to an entirely new science.

7

Cycles, Sunspots, and Pueblos

BEGINNING IN 1901, Douglass devoted an increasing amount of his time and energy to the study of the annual growth rings in trees. From this investigation, he developed the new discipline of dendrochronology, which became the crowning achievement of his scientific career. Originally undertaken to trace solar variations, his extended study eventually proved valuable to the fields of archaeology and meteorology.

Interest in tree rings dated back to the Renaissance. Leonardo da Vinci, for example, made an examination of tree rings and indicated that they reflected age and weather conditions over a period of time. Eighteenth-century naturalists such as George Louis Leclerc de Buffon and Carl Linnaeus employed tree rings to determine the age of trees by the simple technique of counting individual rings. By the late nineteenth century, scientists accepted that normal tree growth resulted in the formation of fairly distinct annual rings of cambial wood. In 1897, a Forest Service official summarized the evidence to support the belief in annual tree-ring growth. B.E. Fernow emphasized that yearly ring growths had long been noted by European foresters who, during the previous half-century, had made tens of thousands of countings. Recognizing that ring types varied according to species and that questionable rings generally appeared in older trees, Fernow calculated that a simple ring counting could produce at least five percent error in the total age of a tree.[1]

The correspondence between tree rings and age was only one part of the early investigation into annual tree growth. External influences on a tree's growth were equally significant. The Dutch astronomer Jacobus C. Kapteyn (best known for his research on stellar distribution and motion) had studied the relationship between rainfall and oak growth in the Rhineland in 1880–81, but believed his research inadequate to publish. Two years later, a physician in Plattsmouth, Nebraska, conducted a similar investigation on a stand of red maple trees planted in 1871. Publishing two articles in *The Popular Science Monthly,* A.L. Child compared the growth of these trees with meteorological records from the six growing months of spring and summer. His study led him to conclude that while there appeared to be a general relation between high temperature, heavy rainfall, and tree growth, the precise ratio thereof remained unknown. Climatic conditions clearly influenced tree growth, but the precise relationship determining that influence was hidden in the factors forming terrestrial weather.[2]

A.E. Douglass began investigating tree growth during his years in Flagstaff. While traveling in northern Arizona and southern Utah during November and December of 1901, he recognized the close relationship between altitude and rainfall, as evidenced by vegetation. Tree growth varied accordingly, with larger and more numerous trees existing at the higher elevations along his route. Because of the sparse rainfall in Arizona, he attributed the predominant influence in tree growth to precipitation. This relation suggested another link. Because the sun's heat caused oceanic evaporation and the winds that dropped this moisture on the continents, variations in solar activity would clearly have an effect on tree growth. Annual tree rings should, in turn, reflect these variations. Thus, tree rings could be used as an instrument to trace and measure solar variation.[3]

Douglass was aware of the growing amount of research into the structure and activity of the sun. Through measurements of solar radiation, Samuel Pierpont Langley and Charles Greely Abbot at the Smithsonian Astrophysical Observatory indicated the variable nature of the sun.[4] At the same time, research into sunspot cycles had produced interesting theories. English scientists in India analyzed the subcontinent's monsoons and suggested that two pulses of rainfall caused the storms, one pulse corresponding to sunspot maximum and one to sunspot minimum. Abbot also believed the sunspot cycle to be important to the Earth's weather. He conceived that sunspot activity led to changes in solar radiation, which affected weather changes. Because of the complex nature of the sun, further research was needed to produce a conclusive answer to the question of the relation between solar activity and terrestrial weather. The suggestion that increased sunspot activity lowered the earth's temperature was challenged by the eminent American astronomer Simon Newcomb. Newcomb argued that terrestrial temperature changes were so small that it was more logical to attribute such changes to "purely terrestrial causes."[5]

Douglass regarded northern Arizona forests as a potential laboratory for studying these theories. He focused initially on studying trees as a measure of past precipitation. Tree-ring size reflected food supply which, in arid climates such as Arizona, was primarily determined by precipitation. Rings thus pre sented a record of past rainfall. From a cursory examination, Douglass concluded that the yellow pines near Flagstaff would be quite suitable for his work. During the spring and summer, soft and rapidly growing tissues expanded the tree's diameter. When growth slowed in the fall, the cells took on an emaciated, reddish appearance, followed by a hard, pitchy ring when tree growth stopped for the winter. This last ring formed a clear division point in growth. Occasionally there were double rings, caused by spring droughts, but these could be differentiated from annual rings by careful scrutiny.[6]

Douglass collected his first tree specimens in January of 1904. Visiting the log yards of the Arizona Lumber & Timber Company, whose president, T.A. Riordan, was a close personal friend, Douglass made his first measurements on a log in the yard. Standing in the snow, he carefully measured and recorded the width of each ring, using a steel meter rule and a magnifying lens. He obtained twenty-four more specimens by cutting thin sections from the ends of logs or from the tops of stumps. He then examined and measured these sections at his leisure. Over the next two years, Douglass obtained twenty-five more tree sections from the Flagstaff area.[7]

By the time he had collected his last specimens in 1906, Douglass had joined the faculty of the University of Arizona. In analyzing the sections, he selected the best sequence of clear and distinct rings on each tree sample. Along a radial line, he carefully "shaved" a one-half-inch wide path, using a common safety-razor blade attached to a short brass handle. In so doing he removed the rough surface of the wood, allowing the rings to stand out more clearly. Douglass next illuminated the sample from the side with a diffused light. He swabbed the entire sample with kerosene to heighten the contrast between the rings, and measured the width of each ring.[8] Although Douglass had developed a workable technique for analyzing specimens, his goal of centuries-long tree-ring records could not be accomplished until he found a means to combine several samples into a single record.

First Steps Toward Crossdating Tree Rings

Douglass's most significant technical discovery, and his greatest original contribution to tree-ring study, was his technique of crossdating tree-ring specimens. During the early work of 1904, he perceived that in each of the first six tree sections from Riordan's lumberyard, there existed a similar pattern of small rings twenty-one years in from the bark. The presence of this pattern in several specimens suggested that the determination of a tree's age need not depend on the existence of an outside bark. If these patterns reflected tree growth accurately, they could serve as benchmarks for crossdating and determining age. To confirm this idea, Douglass examined a tree stump in Flagstaff which had been cut down some years earlier. He noticed the same pattern of small rings, but only eleven years from the outside. By matching the earlier specimens with the rings on the stump, he concluded that the tree had been cut ten years earlier than his first group. He checked with the owner of the property who assured him that the tree in question had indeed been felled in 1894. In such a manner, Douglass verified crossdating as an accurate method of determining the ages of tree specimens. It also provided him with a tool for extending the tree record into the past.[9]

Douglass published an article on his tree-ring research in the *Monthly Weather Review* for June 1909. After discussing his methods of research, he declared that he had discovered strong evidence of cyclic variations in his data. In the growth of rings, cycles appeared with mean values of 32.8, 21.2, and 11.3 years, the last value coinciding with variations in coast rainfall and temperature. More important, the 11.3-year cycle seemed the most evident, and represented the same periodic variation found in sunspots. For the period beginning in 1863, tree growth, precipitation, temperature, and sunspot activity all clustered around the 11.3-year cycle. The interrelationship of these variables suggested a cause and effect. The sunspot minimum, he wrote, "is accompanied by, and presumably causes, the temperature maximum." This, in turn, produced a precipitation maximum. From his research, terrestrial climate and solar activity seemed to have a close relationship. Despite different views expressed by others involved in similar research, Douglass remained convinced that his research could materially add to man's knowledge of the workings of terrestrial weather.[10]

Unfortunately, increased duties at the University of Arizona occupied Douglass's time for the next two years, but during late 1911 he returned to tree-ring research. To expand his representative sample, he obtained seven more cuttings from Flagstaff trees. These cuttings were triangular wedges from logs or stumps, showing the outer fifty to one hundred rings. Douglass also secured sixty-four V-shaped sections from trees near Prescott, seventy miles to the southwest. Because weather records had been kept in Prescott since 1867, he could establish more firmly the relationship between rainfall and tree growth.

Working in his small laboratory in the university's new science building, Douglass carefully analyzed the Prescott samples. The specimens showed the outer fifty years of growth, corresponding to the area's rainfall records. After measuring the first eighteen sections, Douglass realized that the same succession or pattern of rings kept appearing in each sample. A further examination revealed that certain characteristic rings and patterns could be used as markers to locate the tree chronologically. The rings of 1884 and 1885, for example, almost always appeared wider than their neighbors. The clear red ring of 1896 usually showed a double construction, giving a valuable reference point. In each decade, several obvious details could be found. The characteristic rings allowed Douglass to establish the relative and actual ages of his tree-ring specimens, enabling him to trace the weather record in northern Arizona pines.

After this encouraging success with the cross-identification of the Prescott trees, Douglass turned his attention toward the original Flagstaff sections, nineteen of which remained. Flagstaff specimens cross-identified as easily as those of Prescott, confirming the value of this refined method. He then carried his technique to the crucial test. The Prescott and Flagstaff groups, separated in origin by seventy miles of Arizona landscape, were compared with each other in an attempt to cross-identify the two collections. Douglass succeeded. The Flagstaff specimens could be clearly identified with the Prescott trees. The narrow ring of 1851 stood out in both groups. The compressed rings of 1879–85 also appeared in virtually all of the samples examined. The identification of the age of tree-ring specimens had been firmly established by cross-identification of widely separated trees.[11]

The development of cross-identification, and the resultant chronological organization of tree-ring records, represented only one part of Douglass's research. His belief that solar variations, especially sunspot activity, could explain much of terrestrial weather remained the center of his investigations concerning tree growth. He was not the only scientist working on this problem. Late in 1914, Yale geographer Ellsworth Huntington published "The Solar Hypothesis of Climatic Changes." Huntington challenged the arguments of Simon Newcomb and others that small mean-temperature changes were evidence that solar variation had no significant impact on Earth's weather. He argued that the distribution of temperature was equally important as the changes in mean temperature. The varying temperature distributions produced cyclonic storms in the earth's atmosphere, storms which showed a marked correlation with sunspot activity. Using data from 1883 to 1912, Huntington calculated the correlation coefficients (measures of the interdependence between two variables) of sunspot activity and North American storms. He obtained positive values of more than 0.4, a value of 1.0 indicating a perfect positive correlation. From experiments by others, Huntington knew that solar radiation increased with higher sunspot numbers, even though periods of intense sunspot activity produced lower mean temperatures on Earth. His "cyclonic" hypothesis, however, accounted for this seeming contradiction by positing that the increased radiation caused an increased number of storms. These cyclonic disturbances redistributed air masses from low to high altitudes, causing a decrease in terrestrial temperatures.[12]

Although not directly interested in the actual mechanisms of weather phenomena, Douglass nonetheless gained valuable support for his research into solar-terrestrial relationships from theories such as Huntington's. If solar variations indeed had so profound an effect on Earth's climate, Douglass's tree-ring investigations clearly represented a valuable addition to the study of terrestrial weather. The sunspot cycle, so important to Douglass's work, seemed to have a confirmed and significant effect on weather. The discovery of this cycle in a wide variety of trees would show further this interrelationship, with important results for the study of weather.[13]

During his research into the sun's influence on terrestrial climate, Douglass pressed his investigation of tree growth. In this, too, his friend Ellsworth Huntington indirectly contributed to the Arizonan's work. While developing his theory of climatic cycles, Huntington had discovered Douglass' 1909 *Monthly Weather Review* article, which provided a means to test supposed climatic conditions in the past. The Yale geographer had spent several years checking records in Asia and North America of abandoned settlements and droughts in an effort to determine climatic cycles in the historic and prehistoric past. Douglass's article suggested another method to trace cycles: the growth of trees. The sequoias of central California had long been recognized as ancient trees, so the analysis of their rings possibly could provide evidence of cycles as reflected in rainfall. During the early summer of 1911, Huntington measured more than 200 tree stumps in the Sierras east of Porterville, forty of which presented records more than 2000 years old. The general precipitation curves from these trees approximated the curve produced from Huntington's archaeological and historical work in Asia. As he had based the latter research on non-quantifiable sources, however, Huntington could only make suggestions as to the existence of long-term climatic cycles.[14]

Douglass, on the other hand, sought to construct a theory of climatic cycles based on more scientific evidence than the record of droughts in the distant past. These drought records failed to provide information on the severity of climatic changes because they rarely included exact measurements of rainfall. In the fall of 1912, while on sabbatical, Douglass sought additional tree specimens to supply the desired information. He arranged to obtain sections from trees near San Francisco and Portland, as well as from South Dakota, northern New Mexico, and Flagstaff. During a four-month stay in Europe, Douglass also collected tree specimens. From October through January, he obtained samples in England, Germany, Austria, Norway and Sweden. Before leaving England for the United States in late January of 1913, he shipped two boxes of samples weighing a total of 188 pounds to New York and arranged for contacts in Europe to send additional specimens.[15]

Douglass's work depended on establishing a long record of tree-ring specimens. Of the trees first gathered in Flagstaff, two were more than 500 years old. Skeptical of the accuracy of a record based on two samples, he made detailed comparisons of these trees with other trees, focusing on the years they shared in common. Basing his conclusion on the characteristic patterns prevalent in the sixteenth century, Douglass found that while the two oldest records failed to agree in every particular, no significant variations appeared between 1500 and 1600. Although by no means conclusive, Douglass's studies of the oldest Flagstaff trees extended the chronological record back more than 500 years. But a record of 500 years was an inadequate base upon which to build a theory of climatic variation.[16]

During the summer of 1915, Douglass inspected the same sequoia grove that Huntington had visited four years earlier and obtained fifteen sections from various logging camps in the area around Hume, California. Because tree stumps there were often twenty or more feet in diameter, he could not ship complete cross-sections of the trees to Tucson. Therefore, he hired two assistants from the local logging camps and instructed them to prepare long, triangular sections from the top of the stump. Using a long saw, the men cut a section eight inches wide, extending from the center to the outside and showing all the necessary rings. The samples were then marked for identification and shipped to the University of Arizona, where Douglass carefully measured and analyzed the ring sequences during the next year. By the summer of 1916, he had extended his sequoia sequence back in time 2200 years.[17]

Douglass believed that far older trees existed in California, since Huntington in 1911 had found several specimens estimated to be over 3000 years old. Securing funding of $250 in April 1918 from the American Association for the Advancement of Science, he set about extending the sequoia record back 3000 years. He returned to California in the summer, but had difficulty finding older trees. He went south to the General Grant National Park (now part of King's Canyon National Park) and examined various stumps until finally, in the area known as the "World's Fair District," he found a suitable specimen.

Listed simply as "Number 21" in Douglass's records, this stump proved a gold mine of information. Cut several years before, the exposed surface had become carbonized and brittle. The wood tended to break off frequently, clogging the saw with small pieces of valuable tree-ring material. Each bit of wood had to be recovered, marked, and catalogued before cutting could continue. Eventually, Douglass and his assistants shipped the sample to Tucson, where they reassembled it into a triangular specimen over nine feet long. Upon analyzing the new section, Douglass could provisionally date the oldest definite ring at 1304 B.C. The 3000-year barrier had been successfully broken.[18]

Yet the sequoia record remained imperfect. A questionable ring (1580A) appeared in a few of the samples between the supposed rings of 1580 and 1581, causing a year's uncer tainty in the chronology. During the summer of 1919, therefore,

Douglass once again went to the sequoia forests of central California hoping to solve the mystery. Selecting twelve specimens from the most favorable areas, Douglass ultimately confirmed 1580A as a distinct and true ring put down in a very dry year. The sequoia record now started at 1305 B.C.[19]

The gathering and identification of tree-ring samples occupied only part of Douglass's time. His chief goal remained the discovery of climatological information from studying the growth of trees, information which he hoped would lead to a clear picture of Earth's climate. Douglass's continued careful analysis proved quite rewarding. Especially in the dry-climate tree groups, ring width paralleled the available recent rainfall record with an accuracy of seventy percent. Douglass, however, remained unsatisfied, for he believed that this accuracy could be improved by taking into account the amount of moisture "conserved" from previous precipitation. Rainfall per se did not cause tree growth; moisture did. Working with the Prescott tree samples because of the superior record of rainfall in the region, Douglass laboriously constructed an equation to take into account the conservation of moisture, as well as the decrease in ring size with age. Applying this equation to the samples, he increased the retrodictive accuracy of his tree-ring specimens to eighty-two percent.[20]

In 1917 Douglass pursued a further method of confirming the correlation between tree growth and climate. With his chronology providing a detailed record of tree growth for several hundred years, he compared his series of tree-ring dates with the accounts of weather and crop failures mentioned by the famous historian Hubert H. Bancroft in his *History of Arizona and New Mexico* (1889). Of fifteen such incidents recorded by Bancroft, fourteen corresponded to the tree-ring record. The most dramatic agreements concerned the 1680 Rio Grande flood, the famines of the 1680s, and the Arizona droughts of 1748, 1780, and 1821.[21]

Douglass's "Periodograph"

The search for cycles revealed in the annual growth rings of trees took Douglass beyond the arena of pure science. He soon was involved in a series of technological inventions and innovations designed to simplify the discovery and analysis of cycles and related phenomena. He first began this series of inventions in the spring of 1913 in Boston, after his return from Europe. His idea for the "periodograph" came from the late nineteenth-century work of the British physicist Arthur Schuster, who proposed a representation of variable time quantities in the same manner as radiation wavelengths were shown by a spectrogram. Employing large quantities of data, Schuster charted the intensities of various time values in his data. Douglass, however, perceived that a mechanical equivalent could be constructed and used in his own research. Using facilities at the Harvard Observatory, Douglass began constructing his device with whatever materials he could find, including the workings from his brother Malcolm's "grandfather" clock.[22]

The purpose of Douglass's instrument was to display periodic phenomena by means of a "photographic summation" of the data under analysis, beginning with the sunspot record as a test. He first constructed a "multiple plot," a series of curves one under the other, cut out in white and pasted on a dark background. Each curve was identical, but placed ten years to the left in each successive line. If the data from which the curves were constructed displayed a period of ten years, the periodogram would show these phenomena as a series of vertical crests. If the period were greater or less than ten years, the line of crests would slant to the right or the left, respectively. Douglass produced the periodogram itself on a photographic plate positioned behind a lens arrangement which focused the slowly rotating multiple plot on the film as the plate moved downward behind a slit in the focal plane. The "picture" thus produced tested the existence of cycles of various length. If a cycle existed, the crests and troughs of its curves appeared on the film plate as corrugations. If no cycle existed, a uniform light intensity would pass from the curve through the lens, making a uniform image on the plate. Douglass freely admitted that his device could not measure the amplitude of the cycle precisely, but it clearly showed the existence of a period, a valuable asset in any study of cyclic phenomena.[23] Douglass improved this photographic instrument over the next four years, creating by 1918 a device which automatically tested several different cycle

Multiple plot of sunspot numbers, 1755–1911.

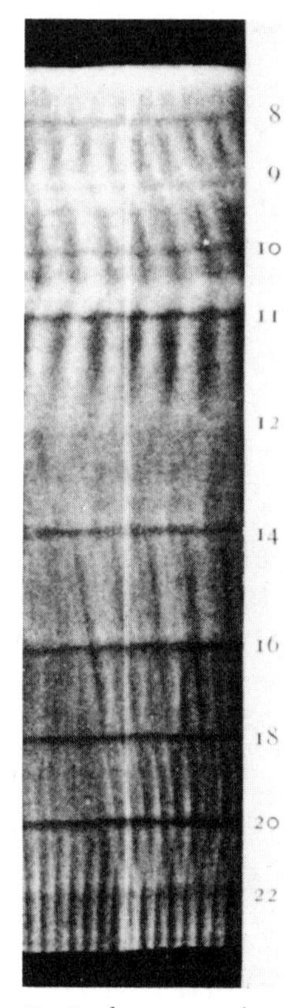

Periodogram of sunspot numbers, 1755–1911. Numbers indicate length of periods shown by corrugations. This was the first periodogram produced by Douglass in 1913.

values. The periodogram resulting from this newest invention showed different periods as differently angled patterns of light and dark streaks, enabling Douglass to record relatively quickly the periods indicated by large collections of data.[24]

The completion of Douglass's newest periodograph paralleled the preparation of his first major publication recounting his research. Written for the Carnegie Institution of Washington, which had been financing part of Douglass's work since mid-1918, *Climatic Cycles and Tree-Growth: A Study of the*

Annual Rings of Trees in Relation to Climate and Solar Activity appeared in print in November of 1919. The first of three volumes (a fourth was never completed), this study presented the history and thought behind Douglass's investigations.[25]

The magnitude of Douglass's research was clearly apparent in his book. Over the preceding fifteen years, the Arizona astronomer had collected some 230 different tree specimens, most from northern Arizona, the Sierras of California, and the Baltic drainage area of Europe. This number, however, presented a misleadingly small indication of the amount of labor performed. Douglass and his assistants had measured more than 75,000 individual rings to gather the desired information from the tree samples.

Douglass believed he possessed ample evidence to support his concept of the interrelation between sunspot activity and the growth of trees. The Flagstaff pines gave early support when he observed that ten of the fourteen sunspot minima and maxima over the past 160 years had been followed four years later by parallel minima and maxima in tree growth. The wet-climate trees of Europe showed this phenomenon even more strongly, with many specimens displaying a precise eleven-year sunspot curve.[26]

The tree-ring evidence led Douglass to conclude that the sunspot cycle had been operating since A.D. 1400 in the same general manner as at present. One notable exception to this operation was an interruption of the cycle in the late seventeenth and early eighteenth century. The precise correlation between solar activity and tree growth remained an open question, but Douglass's research at least made some intriguing suggestions for further work. As *Climatic Cycles and Tree-Growth* entered its final stages of printing, Douglass stepped toward this further work by completing an analysis of the Flagstaff trees, establishing a complete record since 1463 which also cross-identified fairly well with the sequoia record from the same period. It was clear, however, that extending the superior Flagstaff record farther into the past would soon be impossible because of the lack of older trees. Another source for the tree-ring record would have to be found.[27]

Dendrochronology and Archaeology

A few years earlier, Clark Wissler, Curator of Anthropology at the American Museum of Natural History, had suggested a possible source of older tree-ring records. He asked Douglass if it would be possible to treat wood samples from archaeological ruins in Arizona and New Mexico by his method. Wissler hoped to be able to derive the age of the ruins by connecting archaeological wood specimens with those already analyzed by Douglass. The Arizona scientist quickly told Wissler of his enthusiasm for such a project and asked for whatever wood samples the museum would forward. Unfortunately, the museum specimens proved to be in very bad condition and poorly identified, making accurate analysis impossible.[28]

During the summer of 1915, a museum expedition to the Mesa Verde ruins of Colorado discovered a series of log specimens which appeared suitable for Douglass's techniques. Douglass remained interested in the project, but privately expressed doubts about the ultimate success of the endeavor. Writing to his friend Ellsworth Huntington, he expressed hope that old trees from northern Arizona would cross-identify with later tree specimens in the ruins, but confessed that he was "not expecting too much in that line." Wissler arranged for Douglass to receive the tree specimens from the Colorado ruins in the hope that he could develop some sort of chronology. The samples reached Tucson in mid-March of 1916. For the next two years, Douglass attempted to achieve a degree of success with the ruin specimens; but even after a second shipment of wood samples, he possessed too few examples to complete any exacting work.[29]

In the spring of 1919 some of the difficulties involved with specimens from archaeological ruins in the Southwest seemed to clear. The excavation of the Aztec ruins in northwestern New Mexico provided Douglass with a number of very good specimens. By May, he had calculated a provisional date for the ruins which seemed much older than the archaeologists of the American Museum thought possible. Not sure which date was in error, Douglass kept his results private until more research could be accomplished. During early August, he visited the Aztec ruins and obtained a good series of cuttings from a stand of pines growing thirty miles to the north. The famous archaeologist Earl H. Morris guided Douglass around the excavation to acquaint the visitor with the museum's research and solicit recommendations for the further collection of wood specimens. By November, Douglass had analyzed a number of the Aztec beams and was attempting to develop a sequoia record from the best California specimens to compare with the Aztec wood samples.[30] Data from the region's ruins appeared to represent an important addition to Douglass's continuing research and once again emphasized the significance of the Southwest to Douglass's unique scientific contributions.

The American Southwest had long been recognized as an area rich in archaeological data. Spanish explorers in the sixteenth century included descriptions of ruins in their accounts, as did their American counterparts three centuries later. Beginning in the 1880s, serious archaeologists such as Adolph Bandelier, Frank Hamilton Cushing, and Jesse Walter Fewkes penetrated the region. Surveying Arizona and New Mexico for prehistoric ruins, they discovered many sites, including Casa Grande, Pueblo Bonito, and the cliff dwellings in Canyon de Chelly. Excavations began shortly after the turn of the century, when these archaeologists were joined by promising scholars such as Alfred V. Kidder and Edgar L. Hewett. Major excavations were undertaken at Casa Grande, near Florence, Arizona, in 1906 and at Mesa Verde, Colorado, in 1908 and again in 1915. For the most part, archaeologists viewed the Southwest before 1920 as a source of new data, the collection of which became their most important function. Douglass's preliminary attempts to employ tree-ring data to establish the dates of ruins accompanied the opening of the Southwest to intensive archaeological analysis.[31]

Douglass's trip to the Aztec ruins in New Mexico in August of 1919 led to a significant change in the collection of archaeological beam specimens. The Aztec sections previously examined had come from a pile of loose timbers, giving no indication of their original source. Douglass recognized that the location of the samples in the ruins would provide important information both for his own work with tree-ring chronologies and for archaeologists' desire to date the site. To this end, Douglass and his colleagues began using an instrument which bored into prehistoric beams *in situ* and obtained tubular cores one inch in diameter. These samples provided accurate ring records without sacrificing the structural integrity of the ruin. With the assistance of Morris, Douglass obtained some fifty log and beam specimens from the Aztec site. In addition, his assistants gathered samples from living trees in nearby areas to aid Douglass in establishing a crossdated sequence to date the abandoned pueblo.[32]

Because the Aztec rings failed to cross-identify with the living trees of the vicinity, Douglass could not establish precise dates. Yet the Aztec rings appeared to extend for almost two centuries. Douglass therefore compared the thirty-seven beam samples (all but five accurately located in the ruins), searching for a logical way to interpret the ring records. Upon examining the patterns, he recognized a large ring in the outer regions of many beams, indicating a year of excessive rainfall. Douglass believed that a relative chronology could be built around this wide ring. Although such a calendar would not precisely date the Aztec ruins, it would provide a framework upon which the beam specimens could be arranged. Douglass assigned the arbitrary date R.D. (Relative Date) 500 to this benchmark ring, and proceeded to construct a chronology, determining that the principal cutting years for the beams were R.D. 524–25 and R.D. 528.[33] Thus, the Aztec sequence of some two centuries possessed a chronology all its own.

Douglass had access to another group of prehistoric beams from the Southwest, again through the courtesy of Clark Wissler and the American Museum of Natural History. During the four summers beginning in 1896, the Museum's Curator of Anthropology, F.W. Putnam, had made preliminary excavations at Pueblo Bonito, a large pueblo ruin some fifty miles southeast of Aztec in northwestern New Mexico. Several beams from this expedition had been deposited in the New York museum. Douglass obtained sections from seven of these and examined them. The outer rings of the Pueblo Bonito specimens showed striking similarities with those from Aztec, allowing Douglass to cross-identify the two groups. This process provided a 250-year record and indicated that Aztec construction took place forty to forty-five years after Pueblo Bonito, as indicated by the beams' cutting dates.[34] Although he had produced no actual date for either pueblo, Douglass's technique of cross-identification proved to be an effective method of chronologically ordering prehistoric ruins.

Hopes for extending the tree-ring chronology received a reassuring nudge in 1921 when the National Geographic Society explored the archaeological ruins of Chaco Canyon, New Mexico. The leader of the summer expedition, Neil M. Judd of the Smithsonian Institution, knew of Douglass's work and described to him at length the society's plans. He also invited him to make personal visits to the Chaco Canyon area as a guest of the society. Although Douglass wished to take advantage of Judd's offer, he had to stay in Tucson to oversee the completion of the new Steward telescope. In fact, he did not even send the tools for taking core samples until the end of August, less than two weeks before the Pueblo Bonito camp closed. Judd managed to obtain three borings from beams at the site, which he stored at the pueblo. Because the National Geographic Society's expedition to Chaco Canyon would span several summers, neither Douglass nor Judd viewed 1921 as a failure.[35]

Despite the seemingly slow progress in obtaining specimens, Douglass's methods clearly represented a valuable tool for archaeologists. This became apparent during his visit to the Museum of the American Indian in New York City in late April of 1922. George Gustav Heye, the museum's founder and director, and Dr. Frederick Webb Hodge, the famous anthropologist, showed Douglass two beams marked from Pueblo Bonito. Upon examining one of the beams, the Arizona scientist recognized that the cutting date was R.D. 497, the same date for ruins at the Peñasco Blanco ruins in Chaco Canyon. Because other Pueblo Bonito beams failed to correspond to this Peñasco date, Douglass questioned the labeling of this beam, which had been donated by the American Museum of Natural History. When the staff assured him that the label was correct, Douglass suggested that the American Museum had made a mistake. A search of the Museum's records proved its origin. It had, indeed, come from Peñasco, as Douglass had stated.[36]

Douglass's ability to date beams within a year, as shown by his demonstation at the Museum of the American Indian, was a sharp contrast with existing archaeological dating. For example, Neil Judd reported on the first summer of his work at Pueblo Bonito in the March, 1922, issue of *National Geographic*.

Examining pottery samples, stratified remains, and other similar evidence, Judd estimated that the Pueblo Bonito complex had been erected sometime between 800 and 1200 years earlier.[37] A clearer time frame was sorely needed, and Douglass's research seemed to point the way.

Equipped with Douglass's borer and his own archaeological experience, Judd began collecting specimens as soon as he arrived at Pueblo Bonito in early June 1922. Among his shipments to Tucson were a number of charred beam sections. While not expecting such material, Douglass closely examined the charcoal and discovered that the annuals rings showed almost as well in charred sections as in unburned wood. In late July, he reported to Judd that the charred beams provided valuable information. One-third of these samples gave a "pretty exact" cutting date, while another third established a general period.[38]

As Douglass analyzed the various specimens, he attempted to associate them with the relative dating sequence completed for the Aztec ruin two years before. More of the thirty-seven Aztec beams had been cut between R.D. 523 and 531, some fifty years later than the beams obtained from Pueblo Bonito by the Hyde Expedition of the late 1890s. Judd's specimens had been cut between these two periods, most of them showing outer rings at R.D. 491 to 495, although one existed at R.D. 506. To save time, Douglass asked Judd to prepare a list of beam sources in the ruin. This list would enable Douglass to anticipate the general age of beams, and would also hasten the identification of cutting dates. Judd enthusiastically forwarded the required information.

In September, Douglass arranged his schedule to visit Pueblo Bonito. He knew his own limitations in the study of archaeology and felt that a visit to the dig would provide a greater awareness of the method and difficulties in archaeological exploration. In addition, he wanted to study beam construction and location, and gain an overview of the entire operation. Douglass stayed at Pueblo Bonito only a few days, then hastened back to Tucson to continue the analysis of the specimens from the New Mexico ruin.[39]

By late October, Douglass had analyzed all the samples from Chaco Canyon. The beams showed two distinct eras of building in the canyon. One lasted 250 years and partially coincided with the Aztec record, while the second period ran some 160 years. The latter formed a distinct chronology, but failed to overlap the longer record, and thus gave no clue as to whether it occurred before or after the Pueblo Bonito-Aztec period. Of even greater significance, however, was Douglass's preliminary conclusion that the five major ruins of Chaco Canyon (Pueblo Bonito, Hungo Pavi, Chetro Ketl, Pueblo del Arroyo and Peñasco Blanco) had been constructed virtually simultaneously within a twenty-year period. The earliest date of R.D. 476 came from the Hyde beams from Pueblo Bonito, while the latest cutting date of R.D. 497 was found in the Peñasco Blanco ruin.

Dendrochronology and Climatology

Douglass's interest in the dating of archaeological ruins was only part of his research during the early 1920s. He remained keenly intent on studying the cyclic behavior of terrestrial weather, as recorded in the annual growth rings of trees. While investigating the archaeological applications of dendrochonology, Douglass refused to allow his focus to shift permanently from his ultimate goal, the long-range prediction of Earth's climatic behavior. As 1920 opened, he put a few of his ideas before the Ecological Society of America. The meeting of the society was held in St. Louis at New Year's, with the American Association for the Advancement of Science. The members of the Ecological Society welcomed Douglass's "Evidence of Climatic Effects in the Annual Rings of Trees," and the society published his remarks in that month's *Ecology,* the first issue of the new journal.

In his paper, Douglass summarized the key facet of dendrochronology, the technique of crossdating over wide geographical areas. He emphasized that characteristic rings and patterns could spread over very large areas, enabling accurate dating to be accomplished. The very narrow ring of 1851, for example, had been recorded in California sequoias, Arizona and Colorado pines, and a Douglas fir from Pike's Peak, 750 miles from the sequoias. Similarities in sequoias from the areas of General Grant and Sequoia national parks, fifty miles apart, allowed the construction of a pre-Christian chronology 1000 years long. Characteristic rings included 993 B.C., a very large ring, and 1008 B.C., an unusually narrow one. The use of crossdating techniques

A.E. Douglass obtaining a core sample near the Forestdale ruin, Arizona, 1928.

clearly established its accuracy in the interpretation of the annual growth rings of trees.

To obtain the most accurate record possible from the ancient sequoias, Douglass had developed a new procedure, which he reported to the Ecological Society. Selection of the most useful sequoia records depended on their measured response to climatic changes. Douglass called this response a tree's "mean sensitivity," and calculated an index by dividing the difference in width of each two successive rings by their mean width. He then grouped these values in appropriate periods to provide a listing of a tree's responsiveness to climate. Trees which possessed low values, indicating little correlation between ring size and weather, were listed as "complacent," while those showing high values were said to be "sensitive." The yellow pines of the Prescott region, for example, showed a mean sensitivity of 0.64, one of the highest values found. The editor of *Ecology* recognized the significance of Douglass's ideas, and in an appended note emphasized that Douglass had developed a new viewpoint on tree growth. The new method tacitly recognized that, in many species, factors other than seasonal rainfall affected ring width.[40] This recognition placed dendrochronology on a much firmer sylvan base.

The continued expansion of Douglass's tree-ring collection was accompanied by the first use of a new instrument during 1920. G.A. Pearson, of the Forest Service's Fort Valley Experiment Station north of Flagstaff, loaned Douglass an increment borer in the early summer. This instrument (a more delicate version of the tool used to obtain cores from ruins) had been designed to sample the growth of living trees in reforested areas, but its application to dendrochronology did not escape notice. The increment borer worked on the principle of an auger, returning a core of four to five millimeters in diameter from the tree samples. Working especially well in the soft wood of pines and firs, the borer could give any length core up to ten inches, which usually provided a century's record of tree growth. The cores were then placed in paper bags, carefully recorded, and taken or sent to Douglass's Tucson lab. Douglass and his assistants then mounted these cores on specially-prepared strips of half-round wood. The samples were then "shaved" to bring out the rings more clearly. The increment borer made both sampling and storage much simpler.[41]

Douglass took advantage of all opportunities to expand his collection of tree-ring records. In mid-January of 1921, the National Electric Light Association in California invited Douglass to present a paper on February 18 at a weather forecasting conference to be held in San Francisco. He gladly accepted this invitation and quickly wrote to friends in the region asking for the location of good redwood groves nearby. The redwood's propensity for long life suggested it would be a very valuable addition to his research.

Two days after reading his paper in San Francisco, Douglass was gathering radial sections from redwood stumps fifteen miles north of Santa Cruz, with the assistance of R.E. Burton, a science teacher at the local high school. The potential value of the eight specimens never materialized, however, as the ring growth appeared very erratic. Even after months of study, no satisfactory cross-identification could be found. Trees ten feet apart cross-identified with each other fairly well, but a tree fifty yards beyond these displayed a completely different record. The California redwoods gave Douglass little of value.[42]

Douglass's search for a clear explanation of terrestrial weather continued to focus on the connection between solar activity and tree growth. The year 1922 brought exciting confirmation of Douglass's belief in the value of tree rings for the study of sunspot activity. One puzzling aspect of his chronological record of sequoias and yellow pines had been the strong evidence of the eleven-year sunspot cycle, which appeared in these two tree species. The trees showed this cycle reasonably well in all periods, except for a prolonged absence of such evidence between the 1650s and the 1720s. In *Climatic Cycles and Tree-Growth,* Douglass could only call attention to this unusual flattening of the sunspot curve, supplying no better explanation than "some possible interference for a considerable interval about the end of the seventeenth century."

In late February of 1922, however, Douglass obtained the solution to the puzzle. E. Walter Maunder, the famous British astronomer, had become aware of Douglass's research and recognized a possible source of valuable information for his work on sunspot activity. Maunder had investigated the recorded occurrence of sunspots since the invention of the telescope in the early seventeenth century, employing historical sources. He found a fairly constant eleven-year cycle, except for the period 1645–1715, when scarcely any sunspots appeared. Maunder asked Douglass if evidence of a similar abnormality had appeared in the latter's climatic research. Maunder closed by observing:

> If there is some evidence of well-defined and long-continued abnormality as to the climate of the time it should give us a direct hint as to the nature of the connection between sunspot activity and climate; if there is no such evidence, then I think we may safely put any such connection between sunspots and climate as an error in inference.

Douglass's enthusiasm for this bit of news may be easily imagined. From a source who had no firsthand knowledge of his work, he had found an important confirmation of his tree ring-sunspot record based on the historical accounts of European observers. The same general period which showed an absence of sunspots also showed an absence of cyclical variation of eleven-year frequency in the tree rings of Arizona and California. Douglass informed Maunder of his own research, concluding:

> It seems to me . . . that you have brought to light a very important corroboration of the relationship between solar activity and terrestrial conditions, for I presume that these tree variations are related directly to the weather.

This newly discovered correlation played an important role in Douglass's April address to the American Philosophical Society. He observed that the Douglass-Maunder correlation "seemed to confirm strongly the idea that the cycles in the trees are not merely real, but they are related to weather elements and cosmic causes." Of even greater significance to Douglass's goal of discovering the cause of terrestrial weather, this new information "gave added weight to the provisional history of solar variation derived from a study of the 3,200 years of sequoia growth.[43]

Douglass found little time during 1923 to work on his climatic records. Increased demands on his time from the new Steward Observatory and an accompanying growth of his class load were primarily responsible for this neglect. By the spring of 1924, however, Douglass's duties became less demanding and he returned to his research, spending the next several months collecting and analyzing specimens from the forests of the Pacific coast. California sequoias consumed most of his attention.[44] Douglass had constructed a chronology for these trees extending back over two thousand years. He needed to combine this record with the Arizona pine chronology to date archaeological ruins and to interpret the larger collection of pine specimens. Douglass sought to develop this cross-identification in the same way he had matched trees of the same species or area. Comparing records, he searched for common deficient years whose patterns would allow cross-identification. Trying to cross-identify California sequoias and Arizona pines posed a problem. In about one-fourth of the Arizona trees investigated, the characteristic deficient year appeared a year later than in the corresponding sequoias. This discrepancy prompted Douglass to make his fourth sequoia trip in July.

In returning to California, he hoped to extend and improve the general sequoia record and find possible indicators to predict the one-year discrepancy between Arizona pines and California sequoias. Having confined his earlier collecting to the southern sequoia forest, Douglass decided to sample the Calaveras grove near Yosemite National Park. This grove had not been logged, making specimens more difficult to obtain because of the lack of stumps. From fallen sequoias, he took fourteen cores which easily dated by comparison with the sequoia records from the more southern groves. In analyzing these specimens later in the summer, he found the oldest tree to be over 1400 years old. Closer analysis, however, showed that the Calaveras sequoias resembled the Arizona pines significantly less than did the sequoias farther south.[45]

At the close of the next academic year, the collection of specimens once again became Douglass's chief concern. During the summer of 1925, he and his wife Ida traveled throughout the West to collect tree specimens. Exploring promising areas in New Mexico, Colorado, Wyoming, Idaho, Oregon, Washington, and British Columbia, Douglass added significantly to his collection. By July 1, they had turned south to Eureka, California, planning to make an extended examination of the redwood groves in the area. Selecting a site near a mill a few miles south of Scotia, he studied carefully a dozen of the stumps. The trees had been of various sizes and were in scattered locations, thus maximizing the chance of finding a valuable record of tree growth. Taking radial sections of the traditional V-cut type, he arranged for their shipment to Tucson. Despite the care in making the collection, these samples displayed no cross-identification, even among themselves. Although the redwood records went back as far as 1000 years, there was no way to use these trees in his research.[46]

The Douglasses spent the remainder of the summer in southern California. With the cooperation of fellow astronomers, he collected twenty-two borings from Mt. Wilson yellow pines which had the typical curve of Sierra Nevada specimens. The best specimen also showed strong Flagstaff characteristics, suggesting a regional similarity for the American Southwest which, if widespread enough, could indicate the interdependence of California and Arizona weather. In early August, Douglass drove to the sequoia forests near Springville, some sixty miles north of Bakersfield. The trees here were far superi or to those of the Calaveras grove examined the previous year, showing growth cycles similar to Arizona pines. Pine specimens from the same area showed typical Sierra Nevada records and displayed a cross-identification with the Springville sequoias.[47] The summer trip had been immensely profitable to Douglass's research.

By the mid-1920s, Douglass was far from alone in his work. A number of scientists, including both botanists and astronomers, were involved in similar research at the time. Professor J. Arthur Harris, head of the botany department at the University of Minnesota, wrote to Douglass in late 1925 about a forthcoming article on the connection between sunspots and Earth's weather. Harris had used the record of sunspot numbers and Douglass's tree-ring measurements to calculate correlation coefficients, and wished to give Douglass an advance report of his results before their appearance in the *Monthly Weather Review* for January 1926. Finding only moderate interdependence, Harris concluded, "Taken as a whole these coefficients indicate a low positive correlation between sun-spot number and tree growth." He added, "The relationship is by no means so intimate as many writers imply."[48]

After Harris's study appeared, Douglass collected his own thoughts on the subject, which appeared in *Science* for March 4, 1927. He stressed the fact that European wet-climate trees showed the sunspot cycle far better than any other specimens and sought to explain this phenomenon. The reason for the greater response in such trees possibly lay in a cause not previously considered, "such, for example, as radiation (possi bly of short wavelength), that is especially favorable to trees growing generally under cloudy skies." This was an interesting idea, but it did little to advance knowledge regarding the relationship between solar and terrestrial activity.[49]

Douglass pressed forward in his research on climatic cycles. Assistance from the University of Arizona took on a more tangible form in the late summer of 1926 when President Marvin gave him space in the basement of the newly completed campus gymnasium for a laboratory. Within a week, Douglass and several assistants had moved tree-ring specimens and instruments from badly crowded quarters in the Science Building and the Observatory into more spacious quarters in the gym. New instruments from Douglass's fertile imagination also were available to aid research on cycles. The measurement of tree rings was greatly speeded by a plotting micrometer which measured rings through a small telescope and recorded on cross-section paper lines proportional to the width of the rings. Also useful was a device called the Longitudinal Plotter, which reproduced ring spacings on recording paper to allow preliminary analyses to be made by laboratory assistants.[50]

Douglass's research into cycles was also regularly supported by the Carnegie Institution. The Second Cycle Conference in December of 1928 focused on the correlation of Douglass's tree-ring work with the research into tree physiology of D.T. MacDougal of the Carnegie staff, who also served as conference chairman. The conference proved instructive to all cycle researchers by presenting current data and results, and was later reported in full in *The Geographical Review.*[51]

The key event of 1928, however, was the publication of Douglass's second volume of *Climatic Cycles and Tree-Growth* by the Carnegie Institution. Douglass had begun work on Volume II in late November of 1923, four years after the publication of the first volume of *Climatic Cycles.* The pressures of other duties (especially those connected with the Steward Observatory), his wish to analyze new specimens, and the refusal of the University of Arizona to grant a leave of absence hindered the composition of the text. A sabbatical leave for the 1926–27 academic year allowed Douglass to complete nearly all of his first draft by early December, and to begin revising it. In mid-April, Douglass forwarded a manuscript to Clements for criticism and suggestions. The manuscript was returned within weeks with very few changes. Clements wrote to John C. Merriam, President of the Carnegie Institution, and recommended immediate publication. Douglass sent a clean draft to the Carnegie headquarters in early June.[52]

Douglass viewed the second volume of *Climatic Cycles and Tree-Growth* as an extension of the first. By 1928, he and his assistants had dated and measured over 175,000 growth rings from trees obtained throughout the world. As had been obvious almost from the beginning, the western yellow pine (*Pinus ponderosa*) proved to be the most useful species for climatological studies. Its wide distribution, precision in ring records, and five-century age established it as the standard tree for dendrochronological research. The ancient sequoias of California proved valuable for their great age, which allowed frequent cross-identification back to circa 1000 B.C., but further study would be necessary before the sequoia and pine records could be interwoven into a single chronology for the entire Southwest. Other tree species also proved of value. Pines in the eastern United States cross-identified with each other almost as well as the yellow pines of Arizona, if carefully selected. Trees such as hardwoods, cedars, and spruce proved of little value because of their complacent ring records, which often resulted from their homogeneously moist environment. The redwoods of the Pacific Coast, despite their great age, provided no usable information because of the species' failure to cross-identify with neighboring trees. Douglass also blamed this fail ure on the moist climatic conditions of the coast region, but considered the close grouping of redwoods a contribu ting cause.[53]

Once usable trees had been selected and sampled, the crucial work of analysis began in Douglass's laboratory at the University of Arizona. The first step, as he described it in his second volume, was to organize the trees into groups of similar characteristics, corresponding to fairly limited geographical areas. For his cycle studies, Douglass chose the best 305 trees from his collection, which gave a total of 52,400 rings dated and measured. He apportioned records into forty-two groups, and in turn combined the groups into three large geographical zones. The Arizona, or interior zone, included fourteen groups, with 104 trees and 21,210 measurements from the desert Southwest of Arizona, New Mexico, southern Colorado, and Utah. The Rocky Mountain, or eastern, zone included the mountainous regions of the northern intermountain West and bordering areas. This zone comprised fifteen groups, 82 trees, and 14,135 measurements. Douglass placed the 119 trees from the Pacific Coast region into thirteen groups and took 17,055 measurements. This category contained the collection of giant sequoia specimens. For each group, Douglass and his assistants made full and partial curves of tree growth, analyzing both minima and maxima in an effort to isolate meaningful cycles of value to climatological studies.[54]

Douglass's conclusions focused on the cycles which appeared in the data. Relying on the empirical record, he found a number of cycles in the various tree groupings. The long Flagstaff record, for example, which extended from 1300 to 1925, showed an encouraging 11.3-year solar period for 600 years, but with an interruption from 1630 to 1850. In addition, phases of seven, fourteen, and twenty-one years appeared in the Flagstaff record beginning in the 1660s, becoming firmly established after 1700. Analyzing drought records in Flagstaff pines, Douglass found a good indication of fourteen and twenty-one year cycles, with major dry periods occurring at 150-year intervals and minor droughts every forty to fifty years. Sequoias showed distinct ten, eleven, and fourteen-year cycles, a very prominent twenty-year pulse and a probable twenty-three year oscillation. These California giants indicated a further supposed sunspot cycle of almost eleven years during several distinct periods from 1300 B.C. into the twentieth century. However, significant gaps occurred in the record in which this cycle failed to appear, particularly from 1100 B.C. to 300 B.C.

The analysis of western trees reflected a wide variation of cycle values, ranging from eight through twenty-three years. Douglass's early belief in the important effect of sunspots on terrestrial weather had indeed been confirmed by the discovery of eleven-year growth periods in many of his tree samples. But the occurrence of cycles of other lengths left him bewildered. It could be possible, Douglass thought, that many solar cycles were going on at once, but he doubted that several "mechanical pulsations" of the sun were occurring at the same time.

Furthermore, cycles did not continue permanently, as shown by the appearance and disappearance of the eleven-year cycle in sequoias. The crucial problem remained the inability to find or postulate a physical mechanism to explain the operation of solar and climatic cycles. Why, for example, did sunspots appear at a fairly regular cycle of approximately eleven years? More importantly, what physical mechanism worked in a cycle of even this much variation? These unknowns prevented sound predictions for any cyclical phenomenon. As Douglass himself wrote, "Until we know the physical cause of cycles we can not say how long a mechanical repetition will last, for it may break down at any time."[55]

Beginning in the summer of 1929, Douglass explored yet another aspect of his work. A distinct and separate laboratory for dendrochronological research had become, in Douglass's view, a necessity. His research on tree rings was scattered to three separate buildings on the university campus, making a coherent research program increasingly difficult. The favored site for this new facility, of course, was the University of Arizona itself, which had already made significant contributions to Douglass's work. Other possible locations included Flagstaff (because of its proximity to important sources of tree-ring material), Pasadena (in recognition of the growing importance of the California Institute of Technology and the Mount Wilson Observatory) and Palo Alto, California (boasting a Carnegie laboratory on the campus of Stanford University). Although preferring to remain in the West, Douglass also considered moving his tree-ring studies to Washington, D.C., to increase his interaction with the Carnegie Institution.

During the last few months of 1929, Douglass wrote to his friends in the scientific community, stating his case and enclosing plans for the new laboratory. To Neil M. Judd he confided, "I am actually blocked in my work this minute by lack of suitable space." During a trip east in December, Douglass conferred with John C. Merriam at the Carnegie headquarters, convincing him of the value of continued work leading to a tree-ring laboratory. Merriam thought Douglass's ambitious plan a good one, but financial support for such an undertaking would be difficult in the uncertain financial world of late 1929.[56] The tree-ring lab would have to wait.

Douglass's trip east also gave him the opportunity to address a number of organizations and societies. Included were the Carnegie Institution, the National Geographic Society and the American Anthropological Association. In his addresses and reports he described his investigations into climatic cycles and his even more dramatic researches into the chronology of prehistoric ruins in the American Southwest. Begun in the belief that ancient wood specimens from archaeological ruins would play a "creditable part" in establishing a theory of climatic change, Douglass's study of the beams in ancient pueblos soon became the best-known aspect of his multifaceted research program.[57] In fact, his archaeological work during the 1920s provided Douglass's most concrete contributions to American science.

8

The Secret of the Southwest

DOUGLASS'S INTENSIVE STUDY of the climatic effects displayed in tree growth represented only one part of his research during the 1920s. Simultaneously with this work he labored to develop an accurate system for dating prehistoric ruins through dendrochronology. His immediate goal was to apply the study of tree rings to archaeology in an attempt to provide valuable information on the ancient inhabitants of the Southwestern pueblos. By carrying his knowledge of the annual growth of trees into the realm of archaeology, he made a highly significant contribution to modern science.

In 1922, Douglass's study of archaeological dating rested on a very tentative base. In addition to the Pueblo Bonito specimens, he had obtained sections from beams at the Pecos and Hawikuh ruins in New Mexico, but additional beams were necessary for the creation of a dependable dendrochronological record. By the end of the year, the National Geographic Society had granted Douglass $7500 for a three-year project to determine the age of Pueblo Bonito and Pueblo del Arroyo. Douglass was required to make progress reports to Neil Judd, supervising the Pueblo Bonito excavation, and to release all information through the society's headquarters in Washington.[1]

By the end of May, 1923, Douglass had completed preparations for the summer work. The "Beam Expedition" would be a subsidiary part of the National Geographic Society's continuing investigation of the Pueblo Bonito ruins under Judd. Douglass's sole function, as far as the society was concerned, centered on the acquisition and analysis of old trees and beams in the Southwest in an attempt to date accurately the Pueblo Bonito ruin. The five-man expedition, under the field direction of J.A. Jeancon, of the State Historical and Natural History Society of Colorado, and O.G. Ricketson, Jr., of the Carnegie Institution, arrived in the Hopi villages in northeastern Arizona in June.

Douglass joined the party for the first ten days, helping to collect twenty-two specimens from the ruins and villages in the Hopi region, which he hoped would extend the record of living trees far enough back to connect with the chronology already established. With these specimens, Douglass returned to Tucson, charging the rest of the party to collect beam samples from various Southwestern sites. By the end of the summer he had received over 100 specimens from Chaco Canyon, Mesa Verde and Canyon de Chelly, as well as from ruins in the Rio Grande Valley. Two beams from the Wupatki ruins north of Flagstaff came from Harold S. Colton of the University of Pennsylvania.[2]

During the summer and into the fall, Douglass continued his study of the material in his collection. Judd visited Tucson in mid-September to inspect the status of Douglass's tree-ring research and make plans for the next summer's expedition. Upon his return to Washington, Judd recommended additional and immediate funding for Douglass's laboratory work during the winter. Gilbert Grosvenor agreed to make the grant.[3] With this money Douglass now sought to integrate the various ring records from Pueblo Bonito and also to organize more completely his growing collection of tree-ring material. Although a chronology seemed to be taking shape, much more research was needed to date the Southwestern ruins.

By May of 1926, Douglass believed that he had to have even more specimens. With a $700 grant from the Research Committee of the National Geographic Society, Douglass journeyed to Flagstaff in late June and began planning a trip to the pueblo area of Arizona and New Mexico. He had secured a long-overdue sabbatical for the 1926–27 academic year and hoped to collect samples from living or recently cut trees to determine their similarity to his Flagstaff specimens. Because the Flagstaff trees were an important part of his established chronology, their identity with living trees in the pueblo area, descendants of the trees used by the prehistoric Indians, would provide a final check on Douglass's dating techniques.

Leaving Flagstaff on September 4, Douglass and his young nephew Malcolm, who served as driver and cook, drove northeast into Navajo country and reached Kayenta twenty-four hours later. Here they met the famous Indian trader John Wetherill. The next day, Wetherill guided the two to the top of Mount Lolomai, where Douglass took specimens from suitable trees. During the next two weeks, Douglass also obtained samples at other sites, including Chinle, Canyon de Chelly, Chaco Canyon, the Petrified Forest, and Oraibi. Specimens from all these locations showed the same Flagstaff series pattern, confirming Douglass's belief that no significant differences existed among sensitive trees in the pueblo area of northeastern Arizona.[4]

During September, Douglass reported to Judd on his preliminary analysis of selected Pueblo Bonito beams. He estimated that the Pueblo Bonito cutting dates ranged from R.D. 446 to R.D. 567. Judd quickly replied, predicting that earlier beams would be found. "One hundred and twenty years is altogether too short a period within which the Bonito timbers were cut," he said. Indeed, by the following spring Douglass had found earlier records which extended the Pueblo Bonito chronology back 325 years to R.D. 242. This chronology allowed Douglass to date other ruins as well. Several Wupatki specimens were clearly identified as paralleling the latest parts of the Pueblo Bonito calendar. Beams from the Casa Blanca ruin in Canyon de Chelly and from the Kinbiniyol and Solomon ruins in northwestern New Mexico also fit into the general Pueblo Bonito period, all of them dating in the R.D. 500s.[5]

Douglass painstakingly compared the 100 specimens from the beam expedition with samples from other ruins in the Southwest. Sixty-nine beams from the Wupatki ruin quickly emerged as the key element in the developing chronology. Douglass divided these specimens into two separate chronologies, which he called "large beam" and "small beam" sequences. He had already tied the large-beam sequence to the Pueblo Bonito record, but the small-beam chronology matched neither the Pueblo Bonito record nor living trees.

The later years of the Wupatki small-beam sequence matched the early years of the Citadel Ruin specimens, obtained ten miles to the north. The short sequence from Ruin J, two miles beyond Citadel, had the same ring patterns as the late Wupatki record. Into this chronology, Douglass fitted three sections from Mummy Cave in Canyon de Chelly and ten beam samples from Mesa Verde. Another relative chronology was thus developed which proved of great value in ordering the area's ruins. The Wupatki pueblo had been built first, followed forty years later by the Mesa Verde settlement in southwestern Colorado. Slightly less than half a century later, the Mummy Cave habitations appeared. His Citadel Dating, however, refused to mesh with either the Pueblo Bonito sequence or the modern tree calendar.[6]

By August of 1927, Douglass had constructed a chronology of southwestern tree growth which was almost complete. Only two gaps of indeterminate length remained. The first, known as "Gap A," stood between the "R.D." sequence found at Pueblo Bonito and the "CD" chronology of the Citadel and associated ruins. The second, called "Gap B," stood between the Citadel sequence and the record from modern trees extending back to circa A.D. 1300.[7]

Bridging the Chronological Gap

The growing interdependence of dendrochronology and archaeology prompted the eminent archaeologist Alfred V. Kidder to invite Douglass to the Pecos Conference held in late August at the Pecos Ruin southeast of Santa Fe. In addition to Douglass and Kidder, the conferees included Neil M. Judd, Alfred L. Kroeber, Harold S. Colton, Byron Cummings, Edgar L. Hewett and other leaders of Southwestern archaeology. During the three-day conference, which began on August 29, the participants spent most of their time trying to develop a systematic classification of the culture periods in the Southwest.

From these discussions emerged the famous Pecos Classification, which remained the basic system as of the early 1980s. Based on styles of pottery, the most distinctive and widespread artifact, the Pecos system arranged the ruins in eight divisions, running from the preagricultural "Basket Maker I" to the modern "Pueblo V."

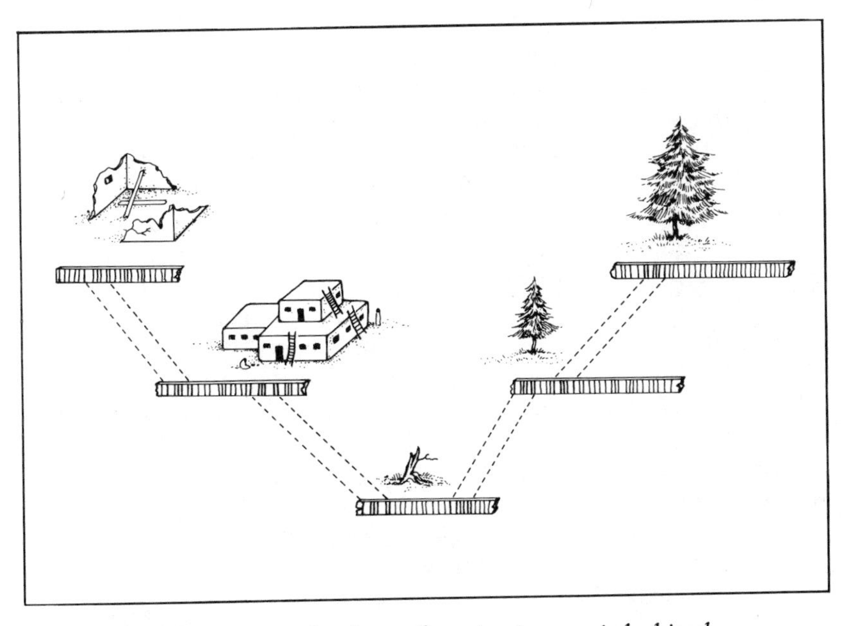

Crossdating samples from (l to r) ruins, an inhabited pueblo, and trees of various ages. Ring patterns ("signatures") provide the key to extend the chronology.

In attempting to attach dates to the new classifications, the scientists turned to Douglass's tree-ring research, believing it the only known method by which actual dating could be made. On the second day of the conference, Douglass presented a brief address entitled "Archaeological Side of Tree Rings and Climatic Records." He explained his technique, showing how the bridge method led to a gradual extension of the tree-ring chronology. He emphasized, however, that the gaps in the record prevented the calculation of absolute dates. To close these gaps, he appealed for continued and increased assistance from field-workers to add to his archaeological wood specimens.[8]

After the conference, Douglass sought to close Gap A in his chronology. Specimens arrived from several excavations, including 185 sections sent by Earl Morris, who was working in the Four Corners country of Arizona and New Mexico. By early December, Douglass finished analyzing Morris's specimens from Sliding Ruin in Canyon de Chelly. He discovered that these sections enabled him to place two more Pueblo Bonito specimens in the chronology, which in turn allowed him to date the Sliding Ruin construction at R.D. 590 to 613. This calculation extended the Relative Dating sequence by seventy-five years and closed Gap A, believed to be the shorter of the two, by that amount. Cross-identification between the RD and Citadel sequences remained impossible, suggesting that Gap A was longer than originally thought.[9]

The field work of the National Geographic Society's Pueblo Bonito Expedition, which had been the precipitant of Douglass's active involvement in archaeology, ended in the early fall of 1927. Analysis and cataloguing of the excavated material would require several years. Both the society and Douglass, however, wanted a rapid completion of the dendrochronological work to provide a framework upon which to place the archaeological evidence. Douglass's crowded laboratory at the University of Arizona quickly became a major focus of archaeology in the Southwest.

In mid-January of 1928, Douglass reexamined his dating of Sliding Ruin. New specimens displayed ring records which showed a distinctive pattern which he had previously assigned to the mid-R.D. 400s. This discovery meant that Sliding Ruin had been constructed during early, rather than late, Pueblo Bonito times.[10] Douglass's insertion of the Sliding Ruin sequence into the early Pueblo Bonito chronology exposed a dating error in beam JPB 116 from Pueblo Bonito, a difficult beam which bore resemblance to patterns at both ends of the RD sequence. Moving this specimen to a period some 250 years earlier shortened the Bonito chronology and led Douglass to reexamine the late RD-early CD ring records. Douglass found in an early Citadel specimen a forty-year ring sequence which overlapped the last forty years of the revised Bonito record. Although based on a single specimen, this overlap united the RD and CD chronologies into a long "floating" sequence of more than 580 years. Gap A had been closed by setting year 0 in the Citadel sequence equal to R.D. 554 from the Pueblo Bonito series.

The closing of Gap A confirmed the value of Douglass's bridge method of building tree-ring chronologies. He realized, however, that his chronology provided only a relative sequence of dates. All concerned recognized the importance of closing Gap B (separating the modern, dated trees from the RD sequence) to determine actual dates for southwestern ruins. Judd and the National Geographic Society allowed Douglass to select the site from which to secure necessary specimens. The Hopi village of Oraibi, the oldest continuously inhabited settlement on the North American continent, seemed to offer the most promise.[11] Tree specimens from the original and later constructions of the village possibly could extend the ring record from the oldest living Flagstaff trees.

In late February, 1928, a group of Hopi Indians from Oraibi, including a tribal chief who lived in the abandoned portion of the ancient village, came to the University of Arizona to put on a dancing exhibition. During their week's stay, Douglass and Byron Cummings, president of the university and a noted archaeologist, secured permission from them to visit Oraibi to collect beam specimens. Cummings assigned Lyndon L. Hargrave, an archaeology student who had already developed a pottery sequence for the Salt River Valley area ruins, to be Douglass's assistant on the Oraibi trip. Leaving in March, Douglass and Hargrave visited the Hopi reservation with Harold S. Colton of the Museum of Northern Arizona. Douglass stayed for six days to become familiar with the site, and then left Hargrave to live with the Hopis for a few weeks to continue his collections.[12]

Douglass began examining the Oraibi specimens as soon as they arrived in Tucson. His first analysis showed two separate chronologies at the site. The first (A) appeared as a series of rings that Douglass thought he had previously dated as part of the 1500s. The second (B) was much clearer, presenting a record in "perfect agreement" with the Flagstaff trees of the 1600s. Despite the supposed close proximity of the two records and the apparent extension of group A into the chronology of living trees (but without any agreement with the Flagstaff record), Douglass at first was unable to reconcile the two groups to one record. After further study he arrived at the crossdating by placing chronology A a full 213 years earlier than his original estimate. With this cross-identification established, a "superb" series existed into the 1300s, increasing the dated chronology more than a century. With other specimens from Oraibi, Douglass extended the modern tree record back to A.D. 1260.[13]

His success reaffirmed Douglass's belief that Oraibi would supply the data needed to "bridge the gap" between modern tree records and the long floating chronology from Pueblo Bonito and Citadel. To maximize the number of specimens, he proposed that Hargrave spend the entire summer at Oraibi, living with the Hopi and securing borings from as many available pueblo timbers as possible. Douglass secured the approval of the Hopi chief Tawa-Guap-Tiwa after presenting him with a small gift.[14]

Douglass sometimes encountered great problems in securing beam specimens from inhabited Indian villages. The cooperation of the inhabitants was not always freely given in the name of science and often was not given at all. Douglass learned this in early 1928 when he attempted to secure beams from a kiva at Walpi, southeast of Oraibi. Because it was a kiva, the religious chamber of the pueblo Indians, he offered to pay double the usual price for each log bored. The kiva priest, however, argued that this particular chamber was a very sacred spot, and the price must be quadrupled. A compromise between the two positions was reached. When his tools arrived, the scientist discovered that members of the village objected to his work and would not allow his party to touch any of the beams in the kiva. Turning to other parts of the pueblo for specimens, he had no luck, in part because the federal government had recently supplied fresh spruce logs for use by the villagers in rebuilding their kivas, and they had cut up and used the old logs and beams for firewood.[15]

Despite the difficulties at Oraibi, Douglass pushed ahead. By summer he had dated 137 Oraibi specimens. Even more important, Douglass had reexamined some of his earlier tree-ring records. A large pine log from Pueblo Bonito, which Judd had forwarded to Tucson in 1922, had never yielded its secrets to Douglass's new science. His recent dating studies, however, provided a much clearer chronology and led to further examination of this log. He had little difficulty assigning a cutting date in the mid-R.D. 400s. The former tree's center was clearly R.D. 113. This extended the Pueblo Bonito sequence by 107 years and allowed Douglass to construct a floating chronology that extended from R.D. 113 to R.D. 699.[16]

By late August, Douglass and Hargrave agreed that searching for gap materials in the Oraibi beams was futile. Their dates were too recent. Even old kiva beams used for floor planks provided no aid in closing the gap between the two chronologies. Analyzing the known sequence of pottery types, Douglass concluded that the Jeddito ruins of Kokopnyama and Kawaiku appeared the most likely sources for the appropriate beam material. Hargrave retrieved a number of specimens from Kawaiku which seemed to confirm this opinion. A large juniper log, although not as reliable as pine, nonetheless bore a strong resemblance to patterns from the early R.D. 600s. Pine fragments found in the same ruin dated from A.D. 1363, indicating that Kawaiku had been inhabited on both sides of the gap.[17]

Within a few months, the Kawaiku ruin provided a significant breakthrough in Douglass's studies. Advised by Judd to concentrate on removing the "question marks" in his ring records, Douglass studied the charcoal specimens from the Kawaiku ruin. He was aided by Earl H. Morris, who in October discovered a fist-sized piece of charcoal at the Kawaiku site. Examining this specimen, Douglass discovered a "perfect" set of rings between 1400 and 1468, confirming and correcting the dates from the smaller pieces of charcoal discovered earlier. He could now construct an entire charcoal sequence for Kawaiku from 1357 to 1495, providing the precise dates for the construction of the now-abandoned pueblo.[18] Kawaiku thus became the first southwestern ruin to be dated by dendrochronology.

The Show Low Ruins and Beam HH-39

During the winter of 1928–29, Douglass traveled to Washington for conferences with officials and fieldworkers of the National

Geographic Society. On January 6, 1929, he talked with Judd and Morris about the most likely sites for specimens which would close the gap in his chronology. Because the most important guide to suitable sites (regardless of their age) was the appearance of charcoal, the three scientists decided to consider the Whipple and Forestdale ruins near Show Low, Arizona, as well as several other sites in the pueblo area of the state. In these conversations Douglass also learned that, because of the slow progress in dating the Pueblo Bonito ruin, the interest of the National Geographic Society in supporting his work was waning. The coming summer might be his last opportunity to close the gap. By spring, Douglass had selected the sites to be examined and had received the $5000 grant for his work from the National Geographic Society. He also selected a staff for the field work, turning to Lyndon L. Hargrave, who had recently taken a position at the Museum of Northern Arizona in Flagstaff, and Emil W. Haury, a promising graduate student in anthropology at the University of Arizona.[19] With two field coordinators, two jobs could be kept going at the same time.

The key element in Douglass's new expedition was pottery. Presenting a predictable and readily observed aspect of cultural development, different types of pottery showed a relative chronological succession which could be employed to provide some hint of the age of the ruins. During late March and April, Douglass and Hargrave traveled to possible sites in Arizona's pueblo region to collect samples of potsherds and find the most promising location for their beam expedition. The discovery of a transient phase of orange pottery, believed to exist contemporaneously with the required beam specimens, eliminated all but four of the twenty ruins under consideration. The most promising sites, those which showed the orange pottery in greatest abundance, were Kin Tiel and Kokopnyama, both north of the Little Colorado River, and the Show Low and Pinedale ruins 100 miles to the south.[20]

The Show Low and Pinedale ruins possessed a clear advantage over those to the north. Both sites were relatively close to a pine forest, which increased the likelihood of finding beams which could be accurately dated. Because Douglass had previously received reports that the Show Low ruin contained charcoal specimens, this site became the first choice of the expedition in early June. Haury and Hargrave therefore set up field headquarters on the site, while Douglass established a laboratory in Flagstaff quarters arranged by Colton.

Douglass's expedition was not the first to investigate the Show Low ruin, then called the Whipple Ruin after the family owning the area. The famous anthropologist A.F. Bandelier had examined the site in April of 1883 during his extensive survey of the Southwest. During the four decades since Bandelier's visit, many changes had occurred at the ruin because of its favorable site for settlement. Three modern houses and sundry other buildings now stood on flattened parts of the site. Many of the original building stones had been used in fences and sheds. The owners, searching for pottery, had turned over much of the ruin and had even built a roof over one of the ruin's rooms for use as a vegetable cellar. The Whipple Ruin looked unpromising as an archaeological site, but the existence of transitional pottery and charcoal specimens outweighed the presence of modern civilization.[21]

On the morning of June 12, Haury and Hargrave began the excavation of the Whipple Ruin. For several days, the archaeologists searched for the desired pieces of charcoal, but found only small slivers and flakes of no use in dating. The main portion of the ruin seemed surprisingly barren. After a week of thorough examination, they began considering the rest of the Whipple site. To the north of a fence separating the built-up portion of the Whipple property, Haury discovered a few undisturbed ruin rooms which had apparently been burned. Not only did the new area show excellent charcoal specimens, but the transitional pottery, known as Little Colorado Polychrome, also appeared. Excavations in this inconspicuous part of the Whipple Ruin began on July 19.

Three days later, Douglass and Judd left Flagstaff in the early morning and arrived at Show Low shortly before noon to examine the site. They immediately hurried to the far north end of the ruin, where a large charcoal beam fragment had just been unearthed. Measuring seven inches in diameter by ten inches in length, the specimen looked like an ordinary round beam whose end had been burned off in the shape of a cone, a fairly common pattern. To protect the beam before removing it from the ruin, Douglass and his coworkers carefully wrapped the

Beam HH-39, the specimen which "closed the gap" in 1929. A workman is shown securing the charcoal fragment for removal and analysis.

By Neil M. Judd © National Geographic Society

specimen in twine. They then painstakingly brushed away the remaining dirt and other material surrounding the log. At the first movement, the log crumbled. Rather than a complete beam specimen, the log was merely a conical shell of charcoal and near charcoal.[22]

Despite its unstable quality, the fragment was very valuable, for the ring record had not been irreparably damaged. Douglass tagged the specimen with the identification number HH-39 (the letters chosen by Haury and Hargrave—whence the initials—to identify specimens from the Third Beam Expedition) and immediately took it to a nearby tool shed for preliminary analysis. Studying the charcoal throughout the afternoon, he dated the innermost ring at A.D. 1237, extending the historical chronology backward more than two decades. As evening approached, Douglass and the rest of his party retired to the village hotel for dinner. They took beam HH-39 with them.

After dinner, Douglass, Judd, Hargrave, and Haury moved into the hotel's living room. Here Douglass seated himself at a small table and continued to study the charcoal specimen. He compared the ring record of HH-39 with the "skeleton plots" made of his two chronologies. These plots greatly simplified the ring records by including only years of serious droughts, plotted on graph paper as straight lines proportional to the droughts' severity. Because narrow tree rings were the most easily identified, the skeleton plot patterns showed clearly any characteristic patterns in beam specimens.

As Douglass continued his analysis, the skeleton plot record showed correlation after correlation with HH-39. The lean years of 1299, 1295, 1288, 1286, 1283 and 1280, as well as three more in the 1270s, all matched the beam's record. He confirmed the core ring at 1237, as he had concluded earlier that day. He then turned to the RD skeleton plot, the record of the floating chronology existing somewhere in time before 1300. As Douglass and his party had hoped, an overlap was quickly seen to exist between HH-39 and the RD chronology. In fact, the Show Low specimen overlapped the late decades of the floating sequence by forty-nine years, the central ring resting on the date R.D. 650.

The correlation between HH-39 and the RD sequence provided a surprising and gratifying discovery. The years 1260 to

1286 had existed in both sequences all along, but previous beam specimens had been too small and questionable to provide convincing evidence. There had, in short, been no gap at all, but, in Douglass's words, "one we had closed without knowing it!" Specimen HH-39 had removed all doubt that the chronology was complete.

That very evening, Douglass correlated many of the relative dates of pueblo ruins to the Christian calendar, using his memory of the RD sequence. Matching his earlier arbitrary date of R.D. 500 with A.D. 1087, he concluded that Pueblo Bonito had been inhabited in the eleventh and early twelfth centuries, as had the other ruins in Chaco Canyon. Mesa Verde, Betatakin, and Keet Seel dated from the mid-thirteenth century.[23] Archaeology had been placed in historical perspective.

Douglass completed the study of HH-39 during the summer of 1929, and quickly moved on to further work. He used additional specimens forwarded by Haury at Pinedale and Hargrave at Kokopnyama and Kin Tiel to bolster those parts of the chronology with defective or unconvincing sections. All the new specimens confirmed the results from HH-39. Douglass now confidently began assigning absolute dates to pueblo ruins. He reported to Neil M. Judd in the fall that cutting dates from the Pueblo Bonito timbers ranged from 919 to 1130; most beams had been cut between 1033 and 1092, indicating the height of Bonito activity. Building at the Aztec ruin seemed concentrated between 1110 and 1121, with Wupatki cutting dates extending more than a century after 1087. Show Low's greatest activity appeared to be between 1174 and 1383, while Oraibi had been continuously built and rebuilt from 1370 until the nineteenth century. Douglass's work had clearly provided archaeologists with extremely significant information.[24]

The National Geographic Society, which had generously funded Douglass's work, believed that the story of his research should be told in the *National Geographic Magazine.* This would be an admirable medium for Douglass to describe in a popular manner the nature and dimensions of his discoveries. In early fall of 1929, Douglass sent a manuscript to Judd, who had agreed to act as initial reviewer. The paper proved a disappointment. It contained too much detail concerning Douglass's background work in dendrochronology and the methods of organizing the tree-ring material. Certainly the account was far too complex and detailed for the audience of *National Geographic.*[25] Douglass worked on revisions for a month, but by late October a large-scale revision had been made by the society's assistant editor, William J. Showalter, allowing rapid publication in the December 1929 issue. Showalter's revision of the manuscript was a wholesale rewriting of the report on dating the Pueblo Bonito ruin. Entire passages had been added for dramatic effect. The most flagrant example, which Douglass later mentioned to the anthropologist A.L. Kroeber, was the imaginative construction of Douglass's analysis of beam HH-39 in the hotel at Show Low:

> The history within that carbonized bit of beam held us spellbound; its significance found us all but speechless; we tried to joke about it, but failed miserably. We felt that here was the tie that would bind our old chronology to our new and bring before us undreamed-of historic horizons.

Admittedly dramatic, this account was the product of editorial imagination rather than historical reconstruction.[26]

Douglass's article, published with the engaging title, "The Secret of the Southwest Solved By Talkative Tree Rings," presented the results of his past decade's work to a wide audience. The completion of Douglass's chronological endeavors was immediately recognized as one of the great moments in American archaeology. As Judd wrote, "Completion of your chronology is by all odds the most important thing that has happened thus far in American prehistory." In the months following the appearance of Douglass's article, laymen and scientists alike congratulated the dendrochronologist. Paul S. Martin, of Chicago's Field Museum of Natural History, told Douglass:

> I have just finished reading your report of your tree ring study, published in the National Geographic. It not only amazes me, but convinces me that your study is the greatest thing in American Archaeology that has ever been done. I feel just as excited over your finds as if I had made them myself! They are truly wonderful. I am grateful as an archaeologist that you happened into this subject, for it puts that much of our study of the Southwest on a sound basis.[27]

The Show Low beam HH-39 provided the information for assigning precise dates to forty-two more ruins almost immediately. This number grew to seventy-five during the first few months of 1930. His precise dating of pueblo construction caused archaeologists and anthropologists to reconsider their views on the development of these civilizations, for Douglass's dates covered a much shorter period of time than archaeologists originally predicted.[28] Cultural development had obviously proceeded at a much more rapid rate than hitherto believed.

Completing the Chronology

Douglass remained at the center of North American archaeology throughout the early 1930s, increasingly taking on the role of consultant. As early as May of 1931, he began to prepare specimens for analysis by others, particularly a young archaeologist named John C. McGregor with the Museum of Northern Arizona. Douglass worked closely with McGregor, teaching him to master the science of tree-ring dating. Within two years, Douglass had removed himself from the analysis of Flagstaff material, except for the final checking of all dates. He modestly, though perhaps inaccurately, disclaimed his importance in such matters in a letter to Museum of Northern Arizona director Harold S. Colton. Sending a list of dates, Douglass wrote:

> I am enclosing a summary of the dating attempts I have made recently on Flagstaff material. These dates are in effect my checking of John's work. That means I regard the Flagstaff area as yours for this early material and the publication of any dates should be by you and I should not be mentioned in connection except as merely checking the dates.[29]

Douglass also began considering at this time the upper Rio Grande Valley of New Mexico. He had attempted for several years to extend his Arizona chronology to this region. In the spring of 1931, he detailed a promising graduate student, William S. Stallings, Jr., to begin constructing a chronology for the upper Rio Grande. In June, Douglass and Stallings traveled to Santa Fe to begin field work, which Stallings continued throughout the summer. By the end of summer of 1932, the project had come under the joint sponsorship of Douglass and Santa Fe's Laboratory of Anthropology. Through his work, Stallings created a Rio Grande chronology that extended back to 1200. Douglass checked the chronology's dates, and found them quite accurate. After collecting samples during the summer of 1933, Stallings extended the record to A.D. 1100, establishing yet another chronology for the Southwest.[30]

Douglass attracted many promising students, one of whom ultimately extended her mentor's work into an entirely new geographical area. Florence M. Hawley had studied with Douglass at the University of Arizona before taking her Ph.D. in anthropology at the University of Chicago in 1934, preparing a dissertation using dendrochronology to analyze the Chetro Ketl ruin in northwestern New Mexico. Soon after Hawley received her doctorate, she traveled south to examine moundbuilder artifacts in the Tennessee and Mississippi valleys, continuing this work after securing a position in the anthropology department at the University of New Mexico. Analyzing the difficult specimens of cedar and other wood from the ancient mounds of Tennessee, she began to develop a chronology for the region, frequently sending her data back to Douglass for checking.[31]

Douglass's many projects at this time included the preparation of a technical report on his Pueblo Bonito work of the previous decade. As originally planned, he was to write a detailed appendix to accompany the archaeological discussion of his coworker Neil M. Judd. The two papers would then be issued as the formal report of the National Geographic Society expeditions. Douglass began drafting his part of the project in early 1930, revising it as his busy schedule would permit. By the spring of 1933, however, Judd wrote that the analysis of Pueblo Bonito material presented such a demanding task that his own written account of the expeditions would not be completed for quite some time. He suggested that Douglass go ahead and publish his report as a separate monograph. While disappointed with Judd's difficulties, Douglass began expanding and revising his earlier work. By the end of July, he had finished a manuscript entitled "Dating Pueblo Bonito" which he mailed to Judd for comments and criticisms.[32]

Judd studied the manuscript for more than four months, ultimately concluding that the draft would require a great deal of reworking before publication could even be considered. The returned copy of Douglass's manuscript bore witness to Judd's careful reading, displaying as much of Judd's handwriting as original typescript. Marginal notations included such statements as "frequent repetition of words" and "too similar to NGS article," all pointing toward Judd's chief criticism: the manuscript lacked the clarity found in Douglass's other writings. Because of their long association, Douglass took Judd's criticisms to heart and devoted the first two months of 1934 to rewriting the entire manuscript. Responding to further suggestions from Judd, Douglass completed final revisions during the summer, forwarding the completed manuscript to the society's editors in late August. By February 1, 1935, "Dating Pueblo Bonito and Other Ruins of the Southwest" had been sent to the printers, and by early June began reaching museums, libraries, and prominent archaeologists throughout the world.[33]

As time permitted, Douglass labored on the archaeological record from the growth rings of trees. Although the success of the 1929 Beam Expedition had allowed the dating of many southwestern ruins, Douglass's tree-ring collection still had a number of short floating chronologies. The most important of these rested on two excellent beam specimens from Mummy Cave, both of which showed a clear and characteristic ring pattern or "signature." Douglass called this floating chronology Early Pueblo Dating (EPD), because of its obvious but as yet undefined age.[34] Throughout the late spring and summer of 1931, Douglass continued to analyze what he considered to be his oldest tree-ring records. One section labeled M200, a twenty-inch beam from Johnson Canyon southeast of Mesa Verde, seemed especially valuable because its rings extended for 350 years. Although he could not crossdate it with any other specimen, Douglass believed that the beam could be quite important in extending his chronology. He also hoped that this M200 sequence, which he called Johnson Canyon Dating (JCD), would ultimately connect with his EPD series.[35]

In the fall Douglass grouped the various floating chronologies into two sequences, each extending more than 300 years and both believed to date before A.D. 700. The two sequences were the previously established EPD and a later series named 179D (after the most important beam in the collection, M179). The later sequence failed to crossdate with the EPD chronology, but did with M200 (or JCD), creating a clear sequence of three centuries. As the pieces of the puzzle slowly fell into place, Douglass told Earl H. Morris that the data still seemed to have two gaps: one between EPD and 179D; and the record separating the late years of the JCD sequence from the dated ruins extending from A.D. 700 forward. He enthusiastically suggested: "All this puts EPD in considerable antiquity. Very likely in the B.C.'s."[36]

Within a month, Douglass's attention was diverted from speculation on floating chronologies to a definite extension of the dated ring sequence. In December, Florence M. Hawley uncovered a superb charcoal beam from the Chetro Ketl ruin in Chaco Canyon. This piece crossdated with the earliest part of the historical sequence, providing a much clearer record of the 700s, a very difficult century in the chronology because of a lack of good specimens. Of even greater importance, the beam extended the dated sequence backward to A.D. 643 and suggested a tie-in with the Johnson Canyon dates near the middle of the M200 beam. If this crossdating were correct, the cutting date of M200 would be ca. A.D. 830, which was much later than archaeologists believed likely for the primitive Pueblo I culture of the Johnson Canyon ruins.[37] Morris believed that A.D. 830 was too late for the termination of the JCD sequence, but wrote Douglass that "if you feel convinced of this finding, I shall alter my notions in accordance."

Douglass would accept no conclusion unless it was firmly established with evidence. For the next two months he reviewed the EPD-179D-JCD sequences in his laboratory at the University of Arizona. The specimens Morris had collected in 1931 from the Lukachukai Mountains in northeastern Arizona were found to fit with "perfect ease" into the JCD sequence, which Douglass continued to believe ended in A.D. 830. The greatest portion of the MLK (Morris Lukachukai) collection ended in the period JCD 150 to 200, some 200 years before the end of the sequence. One specimen, MLK-211, ended significantly earlier at JCD 84. The inner rings of this valuable piece of Douglas-fir charcoal proved of even greater significance. The earliest ninety-three rings gave an exact reproduction of the rings 18 to 110 in the Early Pueblo Dating sequence, clearly showing that the EPD series preceded all other chronologies and contained the oldest specimens in Douglass's collections. Douglass, however, hesitated to make a formal announcement because the crossdating depended on only one specimen.[38]

Throughout the spring and summer of 1933, Douglass examined tree-ring sections from the Four Corners country. Early specimens from Obelisk Cave near Shiprock, New Mexico, allowed him to extend his tentative chronology back to the early first century, near A.D. 10. Further analysis uncovered increased evidence to connect the two major floating chronologies at JCD O-EPD 73, but he had not achieved success in firming up the weak point near A.D. 700, which corresponded to the 230s in the Johnson Canyon sequence.[39]

Douglass confirmed the earliest part of the tree-ring record first. Analysis of beam MV-23, taken from the Step House Ruin at Mesa Verde, paid off in February of 1934, when the interior rings were matched to the JCD sequence and its central rings crossdated with the late EPDs. This discovery removed all doubt concerning the EPD-JCD link and pointed toward the early 400s for the first EPD dates.

These floating chronologies, however, would not provide specific dates until the long EPD-JCD sequence could be tied into the tree-ring record dating from the mid-seventh century. Douglass received three excellent charcoal beam specimens from the Smithsonian Expedition at Allantown, Arizona, in July of 1934. His analysis showed that the beams had been cut in the mid-800s and displayed the JCD signature in the early 600s quite clearly, strengthening the supposed tie in the 700s. In February of the next year, John C. McGregor of the Museum of Northern Arizona sent Douglass another piece of charcoal from the Baker Ranch Ruin north of the San Francisco Peaks. This showed the same JCD chronology and, with the Allantown specimens, firmly connected the floating EPD-JCD chronology to the historical sequence.

Douglass could now begin the important work of translating the EPD and JCD dates in terms of the Christian calendar. With this new chronology, he was also able to extend the record firmly into the first century, using one of the Morris beams collected in 1927. This specimen, whose outer ring dated at A.D. 358, had begun its life in A.D. 93. The record of those years overlapped MLK-152, which had previously established a tentative chronology between A.D. 11 and 239. Douglass carefully scrutinized the 146-year overlap and found a perfect match.[40] He now possessed a tree-ring chronology more than 1900 years in length.

Douglass's extension of the tree-ring record to A.D. 11 capped his contributions to the science of archaeology. As a result of his research, students of the ancient Southwest had a clear guide to the chronological development of civilization in the area. Yet Douglass's archaeological contributions represented an unexpected dividend from his tree-ring studies. Despite his achievements, Douglass remained firmly convinced that his greatest potential contribution lay not in the study of past civilizations, but in the prediction of climates.

9

New Challenges

WHILE SUCCESSFULLY APPLYING THE SCIENCE of dendrochronology to archaeological dating, A.E. Douglass also continued his study of terrestrial climate as reflected in the annual rings of trees. Despite other involvements, he made steady progress in these climatic investigations, gathering significant data in Europe, the Pacific Northwest and the Yellowstone region during the early 1930s. Analysis of this material provided Douglass with further evidence of cyclical fluctuations in climate, as reflected in tree rings, but he realized that further work would be necessary to construct a workable hypothesis of long-term weather behavior.[1]

Despite its tentative nature, Douglass's climatic research continued to keep him before his scientific colleagues. Addressing the National Academy of Sciences in April 1932, Douglass discussed the topic, "Evidences of Cycles in Tree Ring Records." In a succinct summary of his work, he focused on an explanation of the cyclograph and its analytical value. He concluded his presentation by announcing his discovery of a number of cycles in tree-ring records, all of which seemed related to the sunspot period of approximately eleven years.

Douglass's paper quickly drew criticisms. The eminent astronomer Henry Norris Russel commented in general terms on all of the papers, but found Douglass's cycle studies especially troublesome. The number of cycles particularly bothered Russell:

> To disentangle the various cycles from one another and from the other fluctuations, is a task of excessive difficulty, and I fear that even generalized prediction of weather conditions by the extrapolation of empirical cycles derived from observation, if attainable, is far in the future.[2]

The chief difficulty was Douglass's concept of cycles. As an astronomer, Douglass understood the problem, but had not previously realized its magnitude. To most men of science, the term "cycle" referred to a regular repetition of a phenomenon which remained constant through long periods of time. Such phenomena as tides and planetary motions represented these types of cycles. Douglass's use of the term, however, was subtly different. In his analysis of climatological data and tree-ring growth, he had found "cyclic" phenomena which, while not continuous like a sine curve, were nonetheless evident. The discontinuous (or intermittent) nature of Douglass's cycles explained why his cyclograph proved the best analytical method. Mathematical analysis would only work on permanent cycles. As he wrote to H.A. Spoehr, head of the Carnegie Institution's Division of Plant Biology, in June of 1933:

> The comment of astronomers on my cycle work hinges on just one individual point, namely, are cycles permanent or not? The approach of the old astronomer is thru permanent cycles and he simply excludes anything that is not permanent. His definition of cycle, therefore, excludes my work immediately, but nature does not agree with him, for we do have these temporary cycles. They say such temporary cycles are accidents; I say they come in part from the sun, and we can see this resemblance when we admit that cycles are not necessarily permanent. I add that my method of analysis is the only one designed for temporary cycles, and that when solar variations and tree ring variations are expressed in terms of my method the resemblance can be seen. My method of analysis has opened a door that has

> never before been anything but a crack. The picture coming thru is complex and cannot be solved in a moment but there is abundant evidence that it is a real picture. Some day people will realize this thing I am trying to do.[3]

Despite his frustration, Douglass believed that the criticisms of his scientific associates were sincere. He at times felt they were blind, misinformed, or simply unimaginative in their refusal or inability to criticise his work on its own merits, but he never despaired. Douglass accepted that they were good scientists, however afflicted by "frozen officialism" they might be. Given the proper argument and presentation of data, his critics could be convinced.

During the early 1930s, however, Douglass experienced an attack from a completely unexpected source. This attack mystified and troubled him, because it came from people who refused to consider the scientific aspects of his work. During the Great Depression, Arizona politicians viewed research as a luxury, and began pressuring Douglass to justify his university salary by increasing his teaching responsibilities.[4]

Douglass had anticipated the changing tide of research support as early as the fall of 1930, when he asked W.J. Showalter, of the National Geographic Society, for the name of a good press clipping bureau to provide items about his work. As he explained:

> I have had it intimated that some of the people of this State do not think a man in a university is working unless he teaches classes for many hours a day. We have to report on our teaching load. I want to make a substitute to show that if I am not teaching so many classes in a room here on the campus, I am getting something over to a very large class of people interested about the country.

Showalter agreed with Douglass regarding the value of tree-ring research. In fact, he stated that Douglass had done and was doing more than anyone to contribute to the reputation of Arizona and to spread the name of the University of Arizona "to the ends of the earth. . . ."[5]

Douglass's teaching load nonetheless grew during the early 1930s. He had begun teaching a course entitled "Tree Ring

Interpretation" in the spring of 1930 with twelve students and three laboratory assistants. As the pressure for increased teaching grew, he expanded his course offerings to include a directed research class for his more able students such as Florence Hawley and W.S. Stallings. The preparation for these classes and the presentation of formal lectures with numerous visual aids took more time than indicated by Douglass's half-time teaching status. Over the next two years, the university situation became even more discouraging, as the state legislature slashed appropriations by a third. Even though Douglass spent half of his total time in teaching, President Shantz frequently hinted that he should abandon his research in favor of full-time instruction.[6]

By the fall of 1933, Douglass was spending from fourteen to twenty hours per week in classroom and laboratory instruction. His teaching duties increased the following year, when he began a series of astronomy lectures to supplement the university's curriculum. For the two months in which he prepared those lectures he conducted no research.[7] Just as Douglass's future outside of teaching seemed bleakest, events on the East coast suggested a solution to his difficulties.

John C. Merriam of the Carnegie Institution knew of Douglass's problems with the University of Arizona. He was also acutely aware that the maximum return from Carnegie's earlier support could not be expected unless Douglass's research continued. The establishment of a climatic theory on the basis of dendrochronology was one of the key interests of the institution. Accordingly, Merriam spent the summer and fall of 1934 enlisting support for Douglass from among the trustees of the Carnegie Institution. In December, the Carnegie directors agreed to support Douglass, providing the University of Arizona would cooperate. Douglass and President Shantz spent several weeks ironing out the details involved in the arrangement. The university agreed to provide laboratory and office space and would pay Douglass a retaining salary of $600 per year, regardless of the amount of the Carnegie funding.[8] A full-scale research program was finally within Douglass's grasp.

Details of the Carnegie project proved gratifying. The institution would supply Douglass with a fund of $13,900 for 1935. This sum would provide salaries for Douglass and his two assistants, Waldo S. Glock and Edmund Schulman, as well as funds for clerical help and equipment. Merriam emphasized that the Carnegie Institution looked upon this cooperative program as a way for Douglass to complete his climatic work. To this end, Merriam stressed that while archaeology and the extension of Douglass's chronology were important, the focus of his cooperative work should remain on the climatic aspect of dendrochronology.[9]

On February 8, 1935, President Shantz publicly announced the Carnegie grant and the university's cooperation in the program. The news brought congratulations from Harold S. Colton, in Flagstaff, who confided to the university president that he had long believed that Douglass should devote full time to dendrochronological research, rather than to routine class work and administrative matters. The arrangement received national exposure when *Time* published a two-page article on Douglass's work in cycles and tree rings, emphasizing the possibility of long-range weather forecasting. Colleagues wrote congratulatory letters as soon as they read the article. Warner Seely, for example, who had been of great assistance in the construction of the telescope for Steward Observatory, wrote, "The science of astronomy will be the loser by virtue of your special activity in your . . . program, but the realm of human knowledge will benefit and so astronomy ought to be content."[10]

Douglass devoted much of his time during the first two years of the Carnegie program to research and analysis, but the most important project in 1936 was the completion of the third volume of *Climatic Cycles and Tree-Growth*. This work had been one of the provisions of the Carnegie cooperative agreement, and an activity Douglass himself had long desired. Working from an outline prepared during the summer of 1935, he focused on detailed analyses of cyclic phenomena in sunspots, tree-ring records, and geological material. Throughout, Douglass continued his insistence on the value of his cyclograph method of analysis, adding the passage to an early outline: "I protest against the idea that the only solutions of problems are algebraic."[11]

By late November, he had completed the first draft of the volume and began forwarding it to Spoehr and other colleagues for criticism. After making revisions, Douglass sent a copy to the Carnegie Institution. A brief review of the manuscript convinced Merriam and his associates to ask the executive board to underwrite funds to publish 1000 copies. Douglass revised and corrected his work through March of 1936, when he sent "final" copies to F.F. Bunker of the Carnegie press and Charles Greely Abbot, secretary of the Smithsonian Institution. By mid-December, bound copies of *Climatic Cycles and Tree-Growth, Volume III: A Study of Cycles* were distributed to interested students of climatic phenomena.[12]

Douglass's final volume in the Carnegie series was a marked departure from the first two. They had largely dealt with establishing the science and application of dendrochronology. Volume III proceeded from the established science and showed the use of cycle analysis in the study of terrestrial phenomena. The first half of Douglass's book outlined the study of cycles and the climatic records shown in trees. He then analyzed solar and terrestrial records for cyclic phenomena. He emphasized the correlation between sunspots and rainfall as shown in certain trees, and produced records of anticipatory and following phenomena, which proved interesting, if somewhat baffling. Douglass ended his discussion with comments on the use of cycles to establish long-range weather forecasting. He carefully described the techniques to be used, but stopped short of making predictions. "The exact outcome of this [cycle] mozaic [*sic*] is not important," he wrote, "since we have not finished testing our hypothesis; but the advance of technique developed by carrying forward this process is most important."[13]

Although his cooperative program with the Carnegie Institution would continue for another calendar year, Douglass looked ahead to 1937 with concern. From his earliest work with large collections of tree rings, he had recognized the need for separate facilities for his investigations. For many years, however, he had been compelled to work out of his small office in the Steward Observatory and whatever storage space he could find on campus. The situation worsened dramatically in the late 1920s, when the study of pueblo chronology brought nuerous specimens to the Tucson campus for analysis. By the msummer of 1933, although operating in larger quarters in the recently completed baseball stadium, Douglass seriously considered moving his laboratory to another state. The increased teaching load demanded by Depression-minded state legislators had decreased his research time drastically. Douglass toyed with the idea of securing out-of-state private philan thropy, but this proved impossible.[14]

As the final year of Carnegie support began, Douglass recognized that his laboratory operations enjoyed a precarious existence. In February of 1937, he wrote to University President Paul S. Burgess, informing him of his desire to assume emeritus status. Of greater concern to Douglass, however, was the future of his research: "My keen desire is to see some kind of permanence given to this Tree-Ring Work even though the beginning is not elaborate." If the university failed to provide the necessary encouragement for such work, "these valuable bits of knowledge will become associated with other institutions and other parts of the country." As a result of his letter, university officials provided Douglass with four rooms and adjoining storage space on the second floor of the new football stadium south of the observatory.[15] His dendrochronological research was finally concentrated in a single laboratory facility.

Despite his pleasure over increased space, the scientist remained painfully aware that the Carnegie arrangement would quickly end. Throughout October and November of 1937, Douglass besieged the university with plans and suggestions for a permanent, adequately financed laboratory. By the end of the year, Douglass's campaign bore fruit. The Arizona Board of Regents on December 4, 1937, voted to establish a laboratory for tree-ring studies on the University of Arizona campus. Not surprisingly, the regents appointed Douglass director of the new facility, with Emil Haury and Edwin F. Carpenter making up the rest of the staff.[16] Although the university had made a formal commitment to a laboratory for dendrochronology, financing the new department proved more difficult. Douglass attempted to use his contacts in the East to locate possible sources of funding, but had no success. The state of Arizona also proved of little assistance; President Atkinson could offer Douglass only $300 over his salary for directing the laboratory for the 1938–39 fiscal year.

Douglass left Tucson on March 14, 1938, for a three-week eastern fundraising trip. In addition to conferring with scientists involved or interested in tree-ring studies, he made direct appeals for financial assistance to the Rockefeller Foundation, Carnegie Institution, U.S. Weather Bureau and Russell Sage Foundation. Douglass also spoke with Carnegie officials about the equipment purchased during the cooperative arrangement with the University of Arizona. He and his colleagues wanted the institution to donate or sell its equipment to the university. In his March meeting with Merriam, Douglass emphasized that the greater part of the material's value would be lost if it left Tucson. As in all his other meetings, though, no decision could be made until other members of the organization had been consulted. The Arizona scientist left the East in early April with no sure knowledge of his effectiveness.[17]

Douglass's return on April 10 came in the midst of final preparations for the inauguration of Alfred Atkinson as president of the University of Arizona. As one of the most distinguished members of the university community, Douglass attended the ceremony on the twelfth, despite the fatigue from his recent journey. The program indicated that an honorary degree would be presented after Atkinson's inaugural address, but because of his recent absence, Douglass had no knowledge of the honoree. At the close of Atkinson's address, Byron Cummings, the famous archaeologist and former university president, began his introduction of the degree recipient. Taking his lead from the inaugural itself, Cummings described the University of Arizona as "an expression of the forethought and the wisdom of the long line of devoted men and women who have labored for her upbringing last week, last year, and the decades that stretch back to the beginning 40 and more years ago." He continued his introduction by observing:

> It is proper, therefore, that we pause a moment and pay just tribute to one of these noblemen, a man who has moved among the people of the state as a real citizen for more than 40 years, a man who has spent 32 years on this campus and endeared himself to all of us by his kindly, gentlemanly bearing and his constant devotion to the best interests of the institution.

Cummings finally ended the evening's suspense by introducting the recipient of the honorary degree:

> Dr. Andrew Ellicott Douglass has proved himself a successful teacher, a capable administrator, and a wise investigator. His work and his researches have brought great credit to the University of Arizona. His work as an astronomer and his special investigations on the causes of climatic cycles and the effects of those climatic cycles upon the growth of trees, etc., have brought him a reputation that has spread beyond the United States and the Americas to Europe and the other countries of the Eastern Hemisphere. The Douglass method of determining dates of ruins in the Southwest by a chart of the cycles of growth shown in the rings on trees has attracted worldwide attention and is being put into use in various parts of the world. No man has brought more honor to the University of Arizona than Andrew Ellicott Douglass. The faculty of the University of Arizona has unanimously recommended him for this special honor and it gives me great pleasure to commend him to you for the honorable degree of Doctor of Science.

The seventy-year-old scientist stood and moved toward the podium, where President Atkinson awarded the degree as the audience rose in standing ovation.[18]

For the next year, Douglass continued his attempts to secure funding for his laboratory. Although he persuaded the Carnegie Institution to leave its equipment in Tucson on a continuing loan, he was unable to convince it or the Rockefeller Foundation to provide financial assistance. Neither an appeal for government aid nor a proposal to establish a laboratory for tree-ring research connected to the Arizona State Museum met with any success.[19] The Tree-Ring Laboratory had the barest support.

Although the Tree-Ring Lab represented his crowning achievement of the 1930s, Douglass and his coworkers constantly sought new ways to further the study and application of dendrochronology. Acutely aware that interest in dendrochro nology had spread far beyond his laboratory, Douglass and Harold S. Colton in June of 1934 organized a Tree Ring Conference at the Museum of Northern Arizona in Flagstaff. The fifteen persons who attended the meeting exchanged ideas and research results, and decided that the discipline required its own separate journal. The first issue of the *Tree-Ring Bulletin* appeared as an eight-page pamphlet the following month.[20] By the following year's Tree-Ring Conference in Santa Fe, Douglass and his colleagues had decided that a Tree Ring Society would be a valuable addition to dendrochronology. Rap idly drafting and approving by-laws, the conference elected Douglass president of the new society, a post he held for many years. Although the membership of the Tree Ring Society grew slowly and the *Tree-Ring Bulletin* frequently had difficulty attracting manuscripts, each served to strengthen the new science of dendrochronology.[21]

As the decade of the 1940s dawned, Douglass enjoyed widespread acclaim for his scientific endeavors. Now in his early seventies, he had built a scholarly reputation through his dating of archaeological ruins and his investigations of cyclic phenomena in climatic studies. Financial problems had hindered his work in the previous decade, but that was by no means an unusual situation for scientists in the 1930s. In his chosen discipline of dendrochronology, Douglass stood preeminent. Yet his preeminence exposed him to the challenges and attacks of new theories and competing scientists. A hundred miles north of Douglass's laboratory, the most serious challenge to his work was about to begin, launched by the wealthy archaeologist Harold S. Gladwin.

The Gladwin-Douglass Controversy

Gladwin had established a reputation by the late 1920s when he founded the Gila Pueblo Archaeological Foundation in Globe, Arizona. One of the earliest investigators of the Hohokam Indians of central Arizona, he had done much to broaden anthropological knowledge of the prehistoric inhabitants of the state. In 1930, after hiring Emil Haury from the University of Arizona, Gladwin decided to establish a laboratory for the analysis of his own tree-ring specimens, depending on Haury's expertise. Gladwin developed the "increment plot," a method by which tree rings were recorded in terms of the change they showed from the previous year. Quite similar in purpose to Douglass's "skeleton plot" (which displayed only those years of excessively narrow rings employed as benchmarks for dating), this new method gained the enthusiastic support of the founder of dendrochronology and an invitation to publish a description of the method in the *Tree-Ring Bulletin.* For all his contributions and ideas, however, Gladwin had an unfortunate proclivity toward very selective use of data, a tendency which led in part to Haury's resignation in 1937.[22] In a few years, the fireworks began.

A.E. Douglass examining a sample in the Tree-Ring Lab, 1940.

Acting somewhat like a protective father, Douglass always paid close attention to other workers in dendrochronology. In Gladwin's case, his vigilance bore disturbing fruit. Douglass recognized that Gladwin had been collecting tree specimens with insufficient care to provide valuable dating material. Only certain trees gave superior records, as Douglass had demonstrated as early as 1909, but Gladwin seemed content to collect as many "random" trees as possible, perhaps hoping that they would "average out." Douglass drafted a letter to Gladwin in early January of 1940, pointing out certain weaknesses, but decided not to start a debate until he had more evidence.[23]

Within the next few months, Gladwin published two studies of his own dendrochronology work as numbers twenty-seven and twenty-eight of his *Medallion Papers*, a series of monographs which detailed the work of Gila Pueblo. Gladwin's dissatisfaction with the "Douglass method" of tree-ring dating stemmed from the few dendrochronology classes he had attended during the spring of 1930 at the University of Arizona. He dropped this course work early in the semester, after discovering "that I did not possess the ability or type of mind to achieve dependable results." He had not been able to recognize or use the various patterns which were the key to the Douglass method. Disappointed in the "art" aspect of the science of dendrochronology, Gladwin began in 1932 to devise "objective methods" to measure and record tree-ring widths, in an attempt to make the process more mathematical and, in Gladwin's view, more scientific.[24]

The first activity in any tree-ring work, of course, was the preparation of specimens. Rather than rely on Douglass's slow, painstaking "shaving" of samples with razor blades, Gladwin decided to replace student assistants with electricity. He employed a belt or drum sander to prepare the specimens for analysis, followed by the application of carbon-silicate paper and floor wax to remove any "fuzziness" left on the wood. Gladwin attempted to improve on Douglass's method in the preparation of charcoal as well. Flattening such samples with a light application of a belt sander, Gladwin next placed the charcoal into a dustproof cabinet. He then directed a stream of abrasive-filled air toward the sample to sharpen the ring contrast, attempting to avoid any pitting of the surface. After the charcoal dust and abrasive were blown away, the specimen received a coating of preservative made of equal quantities of cellulose cement and acetone. The specimen, whether wood or charcoal, was then placed on a measuring instrument which employed a moving stage with a micrometer thread to measure the ring width and record it on a rotating calibrated drum.[25]

The techniques and ideas described in *Medallion Paper XXVII*, however, presented only the basic outline of Gladwin's objections to Douglass's method. The more penetrating attack came a few months later in another *Medallion Paper,* when Gladwin attempted to establish a purely mathematical analysis of tree-ring dating. In his introductory remarks, he wrote that he had decided to suggest these new techniques because of "my appreciation of the . . . necessity of an objective technique by which results may be comfirmed or corrected." Once again, he displayed his displeasure with the supposed "artistic" character of Douglass's work:

> Heretofore Dr. Douglass has been a court of last resort for the confirmation of dates. However, if tree-ring analysis is to be elevated to the rank of a science, objective methods must be employed, methods which inspire confidence by fulfilling the requirement of exactness and which make it possible to duplicate results as demanded in other branches of science. I wish therefore to emphasize that the ideas and methods which we are submitting herein are in no way intended to disparage work which has been done, but rather to recognize the value of such work as preliminary to more detailed study.[26]

Douglass's method of cross-dating specimens by patterns of ring widths struck Gladwin as insufficiently mathematical to guarantee the reproducibility required in scientific theories and methods. The "human equation" seemed far too important. To improve the scientific nature of dendrochronology, therefore, Gladwin employed his use of the increment plot to record ring-width changes. But instead of comparing each ring to the preceding and following rings only (which, after all, closely paralleled Douglass's method), Gladwin plotted the ring widths as variations from averages of various time periods. Experimenting with running averages of from two years to one hundred years, Gladwin calculated the percentage of agreement or correlation among compared specimens. A thirty-year average gave the highest percentage, seventy-seven percent. With an accuracy of that magnitude, Gladwin felt confident that the human error inherent in cross-dating could be safely eliminated in favor of correlation coefficients.[27]

Despite the mathematics involved in Gladwin's work, Douglass doubted that correlations would prove any more accurate than the slow building of chronologies through cross-dating. He continued to believe that this latter method produced a clear and precise record of rainfall variations, and was the most accurate calendar available for the dating of archaeological ruins. Douglass's assistant Edmund Schulman later expanded this view by emphasizing that in matching tree-ring curves, one often found several instances of "almost" agreement before the final cross-dating appeared. He concluded that the use of correlation coefficients based on growth curves was "an ambiguous correlation method which when unsupported yields only blind *probabilities* of accuracy."[28]

The Douglass-Gladwin controversy reached an explosive point in 1943, when Gladwin began to challenge earlier archaeological conclusions on the basis of his new methods. Surveying the literature produced by Douglass and his followers, he criticized the supposed loose employment of the word "date" by the earlier dendrochronologists and rejected the published estimates of specimen dates:

> Dates are frequently published as, say, 900 ± 10 A.D. All that can be said for this is that it is a misguided attempt to endow tree-ring dating with an accuracy for which there can be no authority or excuse. In such a case the only evidence worth consideration is that the observer believes the *outside* ring to date at 900 A.D., and that an unknown number of outer rings have been lost.

In short, he concluded, "*The only definite date which can be obtained from tree-rings is an actual cutting date.*"[29]

Gladwin's dissatisfaction stemmed from his examination of the dendrochronological literature of the past decade. Making no allowance for later modifications, he described the wide range of dates given for archaeological sites in the Flagstaff area, emphasizing that even Douglass and his colleagues could not agree on building dates. This discovery led Gladwin to reinvestigate many of the specimens already dated by Douglass and McGregor at the Museum of Northern Arizona. Using his own methods, Gladwin found no significant disagreement with specimens dated after A.D. 1100, but discovered considerable differences between A.D. 700 and 1100.

Gladwin remained undisturbed by the disagreement between his own and Douglass's dates. He stated quite clearly in his reports that he had much greater faith in traditional archaeological evidence than in tree-ring dating. For Gladwin, the range of Flagstaff dates already published (A.D. 784 to 1115) made no sense at all, for there existed no perceptible differences in the architecture of the early and late sites, all composed of pithouse structures. The only difference Gladwin found among the sites was the tree-ring dates assigned to them, leading him to suggest that an error had resulted from a mistaken analysis of ring patterns which appeared to overlap, but did not. Samples from the late ruins dated in the eleventh century, for example, could quite possible present very similar patterns to those in the ninth, which would more closely match Gladwin's view of the chronological evidence of architecture.[30]

Douglass's colleagues responded to this challenge with alarm. McGregor, for example, suggested a confrontation of ideas, writing to Douglass: "It seems to me in this paper he has gone so far that it certainly demands an answer." Edmund Schulman, completing doctoral work at Harvard, was also drawn into the debate when asked by the Peabody Museum to give a seminar on dendrochronology "to try to iron out the Gladwin tangle." After the March, 1944, seminar, Schulman informed Douglass that Donald Scott, a museum staff member, planned to write Gladwin and "try to get the points of difference definitely cleared." This would be fine, he continued, "if the real thing concerned in that quarter was scientific truth."[31]

Gladwin renewed his challenge in December of 1944 with the publication of *Medallion Papers No. XXXII*, an analysis of the Medicine Valley site sixteen miles northeast of Flagstaff. The same problem seemed to appear. Gladwin's analysis of pottery and tree-ring specimens placed the archaeological ruins of the area one or two centuries before the dates obtained by McGregor. Once again, according to Gladwin, dendrochronology had led to confusion, with several factors contributing to the situation. McGregor's Medicine Valley work had been accomplished between 1930 and 1932, "when Douglass's technique of tree-ring dating was regarded as infallible and no one would have dared question the authenticity of a date." This had led, in Gladwin's view, to an uncritical examination of the dating process. Other reasons for the tree-ring problems included insufficient knowledge of pithouse structures to act as a control, the erratic nature of Flagstaff tree growth (which Gladwin based on an analysis of randomly sampled trees in the area), the introduction of a dating system based on estimated dates of supposed rings, and "carelessness in the publication of dates and unwillingness to admit or correct obvious mistakes."

Once again, Gladwin emphasized the necessity of giving primary weight to archaeological evidence rather than blindly relying on dendrochronology. Tree-ring dating was only a tool, he argued, and one which possessed serious flaws. He was presenting his theories and asking difficult questions in order to improve the archaeological profession. "I hope that in so doing," he concluded in 1944, "I shall at least have succeeded in showing some of the difficulties and alternatives inherent in tree-ring analysis and so abolish the idea that tree-ring dating is an exact science."[32]

Douglass and his colleagues soon began discussing a course of action regarding Gladwin's charges. By late 1945, it had become clear that Gladwin had no intention of embracing Douglass's system, leading the three principals at the University of Arizona (Douglass, Schulman, and Haury) to conclude that their Globe colleague "was guilty of gross incompetence." They rejected any specific action, however, believing that a direct answer would accomplish little.

A reply to Gladwin soon emerged from another source. In Flagstaff, Harold S. Colton rapidly completed a manuscript to answer Gladwin's archaeological attacks, which had greatly distressed the staff of the Museum of Northern Arizona. Entitled *The Sinagua,* after an early cultural subdivision in northern Arizona proposed by Colton and rejected by Gladwin, the book appeared in early 1946, published by the Flagstaff museum. Although directed toward a clarification of the area's archaeology, Colton's work went to great lengths to emphasize the accuracy and validity of Douglass's tree-ring dating system, upon which the museum based its conclusions. Colton charged that Gladwin's attack on the various dates cited in the literature was unfounded:

> Lacking historic perspective, Gladwin forgot that the technique of tree ring dating was evolving fast during the

> 1930s, so that the presentation of data in the first part of the decade was quite different from the presentation of the same data toward the end. . . .Comparing data from early in the decade with that from the end, Gladwin paints a picture of utter confusion which does not really exist. Also he seemed not to realize that the conclusions presented in more recent papers often superseded the early ones.[33]

Even if Gladwin had properly criticized Douglass, Colton continued, his analytical methods would remain suspect. Gladwin had studied a very small sample of Flagstaff material, and had not attempted to build a master chronology for the area. Further, his criticisms of the A.D. 900–1000 Flagstaff sequence ignored the laborious attempts by Douglass and McGregor to confirm the series. No dates had been announced until both ends of the difficult series had been firmly tied down. Gladwin appeared unconcerned with such niceties, as Colton gladly pointed out:

> Several years ago Gladwin wrote to McGregor for material covering the series 900 to 1000 A.D., and it was forwarded to him. He did not ask for and did not receive the material that tied in both ends of the series. McGregor suggested to Gladwin. . . that he come to Flagstaff and review all of this material, which included several fine long plank specimens, which dated both ends of the difficult series. This offer Gladwin did not accept.

Colton willingly admitted that mistakes in dating had occurred, through typographical or clerical errors as well as inaccuracies in procedures. Many of these had been corrected, though, and none of them were of sufficient magnitude to have effect on the archaeological work of the Museum of Northern Arizona.[34]

Colton saved his most damaging criticism for last. One of Gladwin's chief objections had focused on the supposed misdating of site N.A. 2002, a pithouse in Medicine Valley. Distrusting McGregor's analysis, Gladwin had redated five specimens from the site, keeping his idea of the area's pottery development ever before him as the control factor. It was here that Gladwin made the crucial mistake that Colton doggedly pursued:

> Of the five specimens, two of McGregor's dated in the 10th Century and three in the 11th Century. Gladwin placed them all in the 9th Century, making the site contemporaneous with Pueblo I, but he did not know that the pottery from the site was late Pueblo II, agreeing with the 11th Century dates of McGregor. Using Gladwin's own test for the validity of a date, i.e., agreement with the archaeological evidence, we believe McGregor's dates to be correct.[35]

Colton's publication provided Douglass with a needed stimulus to answer Gladwin in his own way. In letters to various colleagues, Douglass attempted to isolate Gladwin's errors and show why they resulted in inferior archaeological conclusions. He wrote to Charles Greely Abbot that Gladwin seemed "unable to see that he is reading something else than rings" because of his questionable preparation of specimens. He told Colton in December that Gladwin's refusal to publish photographs of his specimens raised justified suspicions as to his methods. By early 1947, Douglass could find no value in Gladwin's techniques, as he wrote to Colton in late January:

> His [Gladwin's] ring reading is as if he were trying to balance himself on one wooden leg without knowing that it is not a real leg. His abraded surfaces are not giving him annual rings and he doesn't know it. That is not an intent to deceive, it is just ignorance.
>
> You see we have discovered a process which Gladwin has never understood—pathetic isn't it?[36]

Douglass also answered Gladwin's challenges to his method of tree-ring dating in "Precision of Ring Dating in Tree-ring Chronologies," a monograph published by the University of Arizona in February of 1947. Although composing a clear, concise explanation to convince his audience of the inadequacies of other methods, he also analyzed Gladwin's work in terms of accuracy. The three key essentials for precise dating had always been site classification to obtain sensitive trees; proper specimen surfaces for accurate ring reading; and the most important part of the whole process, crossdating. Gladwin ignored all of these.

Crossdating was the procedure which had alienated Gladwin from Douglass's method in the first place, but the other two elements were also important. Gladwin's method of preparing specimens made true ring records impossible to observe, Douglass said, because it made "hardness the criterion of ring identity instead of cell structure, shape, and color, and is wholly untrustworthy." Had Gladwin published photographs of his specimens, this error would long ago have been brought before the public. His dedication to statistics had also led him astray in terms of recognizing sensitive ring records, a necessity for all dendrochronology:

> Then again there has been a feeling that statistics relieve the student of responsibility. That can only be true if the basic data are homogeneous. They are considered to be homogeneous if the quantitative distribution of errors makes the right curve (normal distribution). But trees and their ring records are not homogeneous. Some are good recorders of climate and some are very bad ones.[37]

Despite the successful attacks from Douglass and Colton, Gladwin refused to surrender. Focusing on a single discovery made by Douglass, the drought of 1276–1299, Gladwin began his new attack by questioning the concept of narrow growth rings indicating low precipitation. He repeated the contention made many years before that tree growth depended on several factors in addition to rainfall, citing the work of several dendrologists. To support his point, Gladwin described several beam specimens in his collection which failed to show the drought, claiming that this destroyed Douglass's scheme of pueblo development.[38]

By arguing in this manner, Gladwin ignored the entire basis of Douglass's theory of tree growth in arid or semiarid climates, displaying the same "climatic blindness" that led earlier critics to disregard the possibility of an Arizona pine growing differently from a New York maple. Douglass had never argued that other factors were unimportant, only that in certain climates and locations favorable growing conditions depended chiefly on precipitation. For that reason, the dendrochronologist needed to be very careful of the sites in which he collected trees. As far as dendrochronologists were concerned, the students of tree growth cited by Gladwin had been examining trees in areas of high precipitation, trees which Douglass had much earlier shown failed to record variations in rainfall.

The Globe archaeologist seemed to be on more solid ground when he questioned the use of the 1276–1299 drought expressed in tree rings as an archaeological datum. He wrote in 1947:

> ...I find it difficult to understand how the theory of drought, as indicated by tree-rings, ever came to be applied to Southwestern archaeology. Both Douglass and Schulman have claimed that correlations can be established between tree-growth and *winter* precipitation, but each of them has been explicit in saying that summer rainfall has little, if any, effect upon the growth of trees. Since it would seem that summer rainfall, during crop-growing months, is the only season of moisture which could affect the agricultural well-being of Pueblo peoples, I do not see how narrow tree-rings can be regarded as providing any indication of economic disaster in ancient times.[39]

Although a logical and justified objection to basing a theory of cultural catastrophe on tree-ring evidence, Gladwin's analysis presented no significant challenge to Douglass. Gladwin's assumption of adequate summer rains, after all, was based on no more sound evidence than Douglass's belief in a rainfall shortage. Moreover, even if the summer rains had come in plentitude, the winter drought's effect on the soil and water table could still have resulted in agricultural difficulties. Douglass, however, had no real fear of being proved incorrect concerning a general drought between 1276 and 1299. The characteristic pattern of thin rings had allowed him to date accurately southwestern ruins. He would leave the cultural interpretation to the anthropologists.

Douglass heard no more challenges from Gladwin, whose tree-ring methods had been largely discredited. His own methods had been confirmed and remained one of the chief methods of archaeology. Douglass's most pressing challenge was behind him. He could now pursue the many activities which continued to attract his attention as he approached the beginning of his ninth decade.

10

The Final Quest

During the 1940s, Douglass had many more projects to deal with than his refutation of Gladwin's theories. As World War II ran its course, he continued to collect the raw material needed to establish his theory of climatic cycles in the past. Although he increasingly delegated dendrochronological work to his younger associates and assistants, Douglass remained the guiding figure behind all tree-ring projects. Exhibiting another aspect of his multi-faceted career, he also briefly returned to active work in astronomy.

American participation in World War II involved many in the university community. When Edwin F. Carpenter, director of the Steward Observatory, was appointed a navigation instructor with the U.S. Navy, Douglass agreed to serve as acting director of the observatory. He had remained quite interested in the facility, and continued to believe that the campus site had outlived its astronomical usefulness. Beginning in 1943, therefore, Douglass once again considered the possibility of removing the Steward telescope to a more advantageous site.

Douglass investigated several sites around Tucson. He checked two hills south of town near the San Xavier mission, but lights from a nearby aircraft plant and probable residential and industrial development in the area led to the abandonment of this site. He also looked west of the Tucson Mountains, but roads, water and electricity proved difficult and expensive to obtain. He briefly considered locating the Steward Observatory east of Phoenix or in the northern part of the state, but these ideas were dropped upon learning of the difficulty in securing such land and problems of transportation. By mid-1944, Douglass had inspected eight general areas for the new site of the Steward Observatory, and had selected a location near the town of Oracle, where Lavinia Steward had spent her later years. This area offered the best combination of atmospheric qualities, availability, and control of future development.[1]

As did much of the land near Oracle, the proposed Steward site rested near mineral deposits. To minimize interference, existing mining claims would have to be purchased and future claims blocked. Unfortunately, the legal procedures all hinged upon each other, so that no purposeful action could be taken until Congress enacted special legislation to withdraw the Oracle site from future claims and put the area under university jurisdiction. Until the specific site was selected, though, Congress could not act; and until Douglass and his staff could gain reasonable assurances from the owners of existing mining claims, the specific site could not be determined. For the next five years, Douglass and Carpenter tried to solve these problems. They annually extended their U.S. Forest Service permit and examined the Oracle area until December 31, 1949, when their last extension expired. At this piont the idea was dropped. Financial and legal difficulties had thwarted the attempt to move the Steward Observatory to a superior site. The estimated $40,000 required for the new observatory could not be collected or budgeted, and the legal conflicts in settling fourteen mining claims in the area could not be solved. The telescope remained on campus until April of 1963, when it was moved to a site on the grounds of Kitt Peak National Observatory. Six years later, it was used to make the first optical discovery of a pulsar.[2]

Douglass's attempts to move the Steward Observatory, however, were only a peripheral endeavor; his main concern during the war years remained the study of tree rings and their relation to climate. On the eve of American participation in the

A. E. Douglass in his office in the Laboratory of Tree-Ring Research, January 1941. The Steward Observatory is in the background.

hostilities, Douglass and the Tree-ring Lab at the University of Arizona were called to aid the defense effort. The federal government planned to step up the production of hydroelectric power at Boulder Dam to increase the manufacture of magnesium in southern California. In conjunction with this plan, the Los Angeles Bureau of Power and Light (which managed the dam) and the Scripps Institution of Oceanography had launched a study of Colorado River runoff to determine if the water stored behind the dam could be depleted for this purpose. By July of 1941, the Scripps oceanographers concluded that a full-time dendrochronologist would be of inestimable value in securing some idea of the probable runoff five to ten years in advance. Both Douglass and Schulman accepted an invitation to a September conference in La Jolla, but only Douglass was able to attend.

Douglass's explanation of the science of dendrochronology, and Schulman's experience collecting similar data from the Gila River watershed, convinced the Los Angeles Bureau that tree-ring studies would be quite valuable. They could provide a longer, and hence more accurate, record of past water supply. Roy Martindale, operating engineer for the agency, offered Douglass a contract for $1000 to collect and analyze specimens from the Colorado River drainage area. These data could provide a key element in establishing a runoff history of the river and help predict future water levels. Douglass wrote Schulman, in the East on a collecting trip, about the project and urged him to return as soon as possible. By late September of 1941, the final contract had been signed.[3]

Schulman left Tucson on October 4 with two assistants from the Bureau. For the next three weeks the party traveled extensively in New Mexico, Colorado, and Utah, collecting appropriate specimens for the runoff study. Douglass joined the group in Flagstaff on October 26 and accompanied the expedition up the east side of the Wasatch Mountains to the Grand Tetons in Wyoming. After reaching Wyoming, the party journeyed south along the west slope of the Wasatch Range to the Boulder Dam area, returning to Tucson via Flagstaff on 4 November.[4]

Schulman completed his exhaustive study of more than 250 specimens in mid-May of 1942, and traveled to Los Angeles to present his report in person. He concluded that approximately five severe drought years occurred every century in the Colorado watershed, with one of every five years an extreme or moderate drought. The same general proportion also applied to years of excess moisture. The interval between major minima (those with less than fifty percent of normal rainfall) usually ranged from seventeen to eighteen years, but the tree-ring records showed seven cases in the past six centuries of two minima occurring within no more than three years of one another.

Although he had not been asked to make an exact prediction for the next few years, Schulman presented what prognostic data he had discovered. He told the Boulder Dam management that the average period of excess or deficiency was approximately nine years, "with, however, a wide range of fluctuation." Considering that the period 1923–40 had been characterized by deficient growth, Schulman made a prediction for the next decade which proved reasonably accurate:

> Statistical probabilities thus lead to the estimate that the decade 1941–50 will be one of *generally* excess growth and winter precipitation as compared with the preceding decade and the 500-year norm; this however is merely a statistical probability.[5]

His appetite whetted for more exact and short-range forecasting, Martindale telephoned Douglass with a request for an improved prediction of Colorado River water supply. He particularly desired a three-year warning of floods or droughts, to correspond with the reserve limit of the dam. Schulman left on another collecting trip. For two months, he made an intensive search for specimens in New Mexico, Colorado, and Utah. Douglass, in the meantime, applied his cyclograph skills to a reduction of the remaining raw data.[6]

For the rest of the war, the work of the Tree-Ring Lab revolved around the search for climatic cycles. Schulman traveled to the Pacific Northwest and California to expand the collection of tree-ring specimens to be used in the prediction study. While failing to receive further financial support from the Los Angeles Bureau of Power and Light or from either the Army or the Weather Bureau (which had expressed interest in studies which might lead to useful weather prediction), Douglass managed to attract funding from the American Geographical Society and the American Philosophical Society, both of which helped pay for field expenses and preliminary analysis.[7]

Douglass had early realized that collections from the western United States were only part of the story of the region's weather. Evidence of the larger weather picture was also needed, especially during the summer months when Arizona trees showed no response to rainfall. The nearest source of this information lay in South America. As early as July of 1942, therefore, Douglass began soliciting support for a collecting trip to South America.

The response was disheartening. Friends and colleagues at the Carnegie Institution, the National Geographic Society and the U.S. Weather Bureau neither had funds to dispense, nor had they any idea of possible sources of funds. Charles F. Sarle of the Weather Bureau suggested that if Douglass could construct a program for weather prediction in east Asia and Japan, funds could probably be obtained. Douglass even wrote to Vice President Henry A. Wallace, who had visited the Tree-Ring Lab a few years earlier, with a request for assistance in securing the needed funds. He received no reply. The only money collected by Douglass during his seven-year campaign was a five-dollar contribution from H.J. Bauer of San Diego, president of Southern California Edison, and a $100 check from L.K. Doutrick, Vice President of Central Arizona Light and Power Company in Phoenix.[8]

Finally, however, an organization showed sufficient interest to make a sizeable contribution to Douglass's proposed South American expedition. The American Academy of Arts and Sciences in Boston gave the Tree-Ring Lab $1500 in late 1948 to support the project. With additional support from the University of Arizona, the following November Douglass dispatched Schulman to South America. Concentrating on the Andes, Schulman spent four months collecting suitable tree-ring specimens, returning in late March of 1950. The Argentine Forest Service and Chilean Agricultural Experiment Stations gladly supplied vehicles, guides, and assistants for the expedition. The specimens collected during this South American journey showed excellent crossdating and provided complete justification for the expensive trip, but represented only a preliminary addition to the knowledge of global weather patterns.[9]

Douglass spent the last half of the 1940s supervising the operations and finances of the Tree-Ring Lab. Although still proposing a separate laboratory building (estimated at $50,000), Douglass expanded the lab into a few more rooms in the stadium. In 1947, he suggested a ten-year financing plan to provide an additional $10,000 each year for field work and analysis, but soon realized that the university would not support the facility to that extent.

Nonetheless, the lab did expand after the war, with increased salaries given both Douglass and Schulman. The 1947–48 budget included a salary for an Assistant Dendrochronologist, a post ably filled by Terah L. Smiley who concentrated on the archaeological aspects of the lab's work. By the end of the decade, the Tree-Ring Lab was a clearly established part of the university community, offering instruction as well as research involving astronomy, archaeology, and climatology. In 1949–50 Douglass requested a budget of $11,103, more than double the first post-war budget. The university administration approved all but $297 of the request.[10] Finances were no longer Douglass's chief worry.

Douglass looked ahead into the 1950s with concern, however. He was now eighty-two years old, and his eyes had been increasingly troublesome as a result of cataracts which had begun developing in 1945. Douglass also suffered occasionally from a duodenal ulcer as a result of overwork. Although he remained reasonably healthy, the scientist realized that the years ahead were limited.

Douglass was most troubled by the tremendous amount of work which required his attention. The establishment of a cyclic interpretation of climatic change and the resultant long-range forecasting remained, as it had for almost half a century, his chief goal, the focus of all his work. Toward this end, he had become increasingly involved in the study of possible planetary influences on the sun, which might in turn affect terrestrial weather. This new field of "astroclimatology" required Douglass's utmost endeavors because of his conviction that he alone could establish such a discipline. Having established a laboratory and trained a capable staff, he could now devote himself to what he believed would become his greatest achievement: the analysis and forecasting of terrestrial weather through the new science of astroclimatology.[11]

The Foundation of Astroclimatology

The request in the early 1940s for weather predictions concerning the Colorado River watershed reinforced Douglass's determination to find the means to establish a science of long-range forecasting. He appreciated the difficulties involved in such a project, but remained convinced that the analysis of cycles displayed in tree rings would lead to success. Douglass believed that the most promising field for study was an investigation of the correlation between tree rings and astronomical phenomena, especially sunspots, but the sunspot record presented an incomplete contribution to such a study. Scientists had not yet determined the cause of this solar activity. Without the knowledge of that cause, or at least an idea of the influences which indirectly caused sunspots, any theory involving these markings would be suspect.[12]

By 1940, Douglass's views on sunspots had focused on the influence of planets on the solar sphere. In this, as he freely admitted, he produced no new theories. As early as 1900, British astronomer Ernest W. Brown had proposed that Jupiter and Saturn, when in certain positions, resulted in increased solar activity in the form of sunspots. Discounting attempts to explain sunspots in terms of one planet only, Brown pointed to the tide-raising forces caused by Jupiter when in correct position relative to Saturn. This period of conjunction or opposition, he theorized, increased these forces and produced a disturbance on the solar surface which appeared as sunspots. Graphs of Jupiter's "wave" of tide-raising forces, the planet's orbital eccentricity, and Saturn's path in regard to Jupiter's, overlapped to a degree and showed a reasonably close correspondence to the sunspot curve. The graphs failed to fit precisely, since the sunspot period of 11.1 years exceeded the cycle of influential Jupiter-Saturn combinations of 9.93 years, but visible resemblances were clear. Some of the deviations, thought Brown, could be explained by the influence of the inner planets as well as by other "arbitrary" quantities.[13]

Similar ideas, all of which Douglass filed for future reference, occasionally surfaced during the next few decades. In 1930, for example, Inigo Jones, Director of the Australian Bureau of Seasonal Forecasting, sent Douglass his theory of astronomical effects on weather, suggesting a combination of the four major planets as the principal factor in droughts and floods. As the solar system moved through space, the sun entered a magnetic field which caused sunspots. When the four major planets, which acted as magnets, appeared in front of the sun, they deflected this field and caused a disappearance of sunspots and hence drought periods on the earth. When the four planets remained behind the sun, the full force of the magnetic field streamed onto the solar surface. This caused an increase in sunspots and flood conditions. The more planets in a direct line, the more severe the terrestrial response. Similar suggestions came to Douglass throughout the 1930s, leading him to obtain a set of diagrams used to derive planetary longitudes, which would allow him to find the conjunctions and oppositions of any desired planet.[14] It was an intriguing topic.

Douglass made no sustained effort to pursue sunspot studies until May of 1942, when the Boulder Dam officials requested the development of possible climatic prediction. Having failed to establish precise forecasting through previous methods, he began to examine the possibility of planetary combinations as a cause of solar variations. Methodically trying single planets and combinations of two planets, Douglass found no relation. But in investigating combinations of three planets, he perceived faint resemblances to some of the cycles found in long tree-ring records. This promising development justified further inquiry.

During the summer, Douglass designed a simple orrery (a machine to reproduce the motions of planets) to provide a method of calculating planetary combinations. Employing a series of gears to reproduce planetary motions, Douglass's "interpolator" (so called because it was designed to interpolate positions of planets) allowed him to time the occurrences of supposed favorable combinations. In his first attempts, he followed E.W. Brown's lead and tested Jupiter, Saturn, and the inner planets, but found no positive results. He then turned to the four major planets and plotted their times of conjunction and opposition. At the same time, he carried out a cyclogram analysis of sunspot periodicities, and was struck by a familiar sight:

> A diagonal on the SS [sunspot] cycle cyclogram shows at once an intermittent cycle close to 10.0 years. A careful test gave 9.98 years. The shock of realizing that this was only ½ of 1 percent from the mean "conjop" (conjunction-opposition interval) of the largest two planets, Jupiter (J) and Saturn (S), lasted for a week. I forgot to note the exact date of this. . . . It was strong evidence that the planets could explain the sunspot periodicities and so extend the idea of 1929 which attached climatic cycles to the sun.

Immediate further studies showed a sixty to seventy percent success in correlating astronomical and tree-ring phenomena for the past 1000 years.[15]

By the late summer of 1945, Douglass recorded the results of his analysis of planetary and rainfall data and his attempts to discover the relationship between the two. Preparing an informal account at his summer home in Altadena, California, he listed the major conclusions of his researches into astroclimatology. Periods of maximum rainfall, as shown in tree-ring records, seemed to correspond with the conjunction or opposition of three major planets. The effect was stronger if the planets combined within a range of twenty to thirty degrees, but their influence seemed to remain significant even when the planets spread over a fifty to seventy degree separation. If four planets lined up in a similar position, the effect on the sun, and hence on terrestrial weather, was much stronger, often lasting three to five years. In a few cases, Jupiter and only one of the other major planets combined in correspondence with certain displays of maximum rainfall. Various cycles of tree-ring growth and planetary combinations appeared to match within reasonable limits. In fact, Douglass's research disclosed that only 1.5 rainfall-tree ring maxima per century (out of eighteen per century) "so far in our studies do not fit this planetary pattern with significant closeness."[16]

Douglass also jotted down a possible explanation of the physical causes underlying his planetary hypothesis. He first postulated that the electronic material floating in the sun's atmosphere "may be moved by gravitational or electric forces from the planets, or possibly by great static electric pressure on the sun for which the planets may offer a channel of escape." This action presumably took place on opposite sides of the sun, pointing toward or away from the planets. Terrestrial phenomena were influenced by this escape, possibly to an increased degree when the earth was "in fair line with the planets that build up the exciting forces (whichever kind) on the sun."

The disruptions on the solar surface caused by this phenomenon led to profound effects on Earth. The material forced away from the sun by planetary forces spread throughout the solar system. Earth felt the effect of this radiation, Douglass recorded in his notes, and was influenced by ultraviolet rays, which were "often connected with sunspot activity and electronic charges that strongly affect terrestrial magnetism." Douglass had to stop at this point, a failing he both recognized and resented. "Just how these radiations produce our storm patterns and become expressed in rainfall," he concluded, "is one of the gaps."[17]

A shortcoming in Douglass's work was the planetary interpolator he used for calculating positions of planets. In the interest of time, and because he had intended to study only recent data, he had not included various orbital anomalies in designing the orrery. After the initial success of both the instrument and his theory, he sought to extend his investigations into earlier centuries. In January of 1946, Douglass checked his planetary data to determine the magnitude of the anomalies involved. The errors in planetary positions were less than ten degrees, but one error reached twice that figure, clearly unacceptable for accurate astronomical work. He immediately began designing the necessary corrections for his instrument and oversaw the interpolator's modification during the early summer.[18]

Convinced of the astronomical accuracy of his instrument, Douglass made an empirical analysis of planetary and sunspot material. Rather than delve into the question of causes and effects, he sought first to establish the cycles (or "cyclics," as he usually called them) which were displayed in the data. The sunspot cycle turned out to be quite complicated. Douglass found that the cycle alternated between periods of ten years (9.93) and approximately 13.5 years. The pattern shifted each 85.7 years, which Douglass immediately recognized as the "conjop" interval of Uranus-Neptune. The theoretical calculations of sunspot maxima from these data closely agreed with the observed years of high sunspots, "with the same total interval and the same number of maxima and the same result" for the period since the 1780s.[19]

Douglass also revised his work uniting astronomy and climatology. In an attempt to draw the inner planets into the equation, he modified the interpolator several times to include the motions of Mercury, Venus, and Mars in the production of terrestrial climate. He also studied one of the most intriguing aspects of his research, the two-year "scatter" cycle of alternating wide and narrow rings, which often appeared in tree-ring records and seemed to be caused by juxtapositions of Mercury, Earth, Mars and Jupiter. These planetary combinations, Douglass concluded, caused tidal forces and stimulated sunspots.[20]

Douglass's attempts to correlate his varied data received a dramatic boost in the spring of 1953. Utilizing both interpolator and cyclograph, which Douglass believed paralleled the "timing method used by Kepler in his planetary laws," he simultaneously analyzed solar records, planetary combinations, and tree-ring patterns of recently obtained pines from San Bernardino, California. These diverse data seemed to agree in an intriguing manner. The "revised" sunspot cycles of a ten-year period for approximately forty years, followed by a variable 13.2-year cycle of the same duration, appeared in the southern California trees, and paralleled the Jupiter cycle as reinforced by the closest passage of Mercury to the sun. This planetary combination effected solar surface changes and hence changed terrestrial weather. The coincidence of data seemed most encouraging.[21]

Yet difficulties clearly remained. After a dozen years of astroclimatic research, Douglass had collected a vast amount of analyzed data and intriguing coincidences, but he still felt unsure of the predictive ability of his scientific endeavors. Further, although he had discovered parallels among certain groups of data, significant gaps continued to plague the development of an accurate theory of astronomical influence on terrestrial climate. The twelve-year cycle displayed in certain

Arizona and New Mexico trees, for example, while valuable in other dendrochronological studies, could not be made to fit any significant planetary data.

An even more troublesome difficulty involved the attitudes of other scientists toward his work. The concept of varying cycles appeared as little more than data manipulation to researchers increasingly devoted to precise mathematical analysis and the development of digital computers. The attitude of his professional colleagues led to increasing self-imposed isolation. Throughout this period, Douglass began several letters to Vannevar Bush, the distinguished physicist and president of the Carnegie Institution, but decided to hold back accounts of his work in the expectation of Bush's lack of enthusiasm. Writing to Fred L. Whipple, chairman of the astronomy department at Harvard University in 1954, Douglass admitted the need for further study to both solidify his own research and to provide it with a fair hearing before the scientific world. In one of the few instances of professional correspondence from the period, he discussed his investigations, adding:

> I am wondering if I should have the astronomical part of it looked over by someone. I am wondering if I could have a real astronomer who knows something about planetary orbits and who has a willingness to look at new research, who could perhaps become completely acquainted with all this astronomical phase of tree-ring work. I particularly believe that I want someone who will not be overcome by the newness of all this. . . .[22]

As an evaluation of this nature was impossible to effect (the Tree-Ring Lab's budget could not support such a training program), Douglass continued searching for more pieces of the astroclimatic puzzle. Collecting tree-ring specimens remained important to his work, but he also advanced into theoretical explanations of the observed phenomena. By late 1956, he had developed a clearer model of the mechanisms involved in the operation of planetary influence. As various planets combined in approximate alignment, their gravitational or electromagnetic force (Douglass was not sure which) had a profound influence on the solar surface. This influence produced increased numbers of sunspots which in turn led to a higher level of electromagnetic radiation flowing away from the sun and striking Earth. This phenomenon caused profound changes in rainfall and river flow, as shown in tree rings. Not all the details were know, but Douglass had at least attempted a rational scheme of cause and effect.[23]

During this period of intensive research, Douglass also drafted a proposed fourth volume for his *Climatic Cycles and Tree Growth* series published by the Carnegie Institution. He had early planned such a volume to provide further information on tree-ring records, but by the mid-1950s he had decided that his climatic investigations presented a far more valuable contribution. Douglass began serious work on his manuscript in January of 1953, but active research delayed the composition of the text until October. Within five months, he completed a discussion of the general background of astroclimatology and the basic tenets of planetary influence on weather. Douglass then began constructing his most important exposition, the theory of weather forecasting.

Over the next few years, Douglass frequently mentioned that his manuscript was "almost" or "practically" finished, but he kept uncovering new or different data which required revisions in his theory. Despite continued modifications, he sought a publisher for his work. In early 1956, Douglass began a series of several letters, none of which ever left his desk, asking if the Carnegie Institution would be interested in publishing a sequel to his series. Even if they proved uninterested, Douglass still wished to make his new manuscript a continuation of his Carnegie work, and requested suggestions on how best to insure continuity. By May, he had evidently reconsidered his early wish for Carnegie publication. He expected a delay if Carnegie agreed to publish the work, and felt this was too risky because of his "deteriorating eyesight." For the time being, he decided to continue his research and further revise his manuscript.[24]

Douglass's interest in astroclimatic research and writing paralleled his duties as director of the Tree-Ring Lab. Although he had long wished to establish Tucson as the central data bank for tree-ring studies, he had been unable to accomplish even the beginnings of this goal until 1950. In January of that year, Harold S. Colton wrote that the Museum of Northern Arizona could no longer afford a full-time curator to take care of their

5300 tree-ring specimens. He felt that the University of Arizona was the most logical place for their deposit and asked Douglass if such a transfer could be arranged. During the next few weeks, Douglass, Colton, and University President J. Byron McCormick worked out the details of the proposal. The Museum of Northern Arizona relinquished its collection as a permanent loan, with the stipulation that the specimens be organized and available to the public. The agreement also provided that the Tree-Ring Lab staff would date new material collected by the Flagstaff museum. To guard against later problems, Colton further insisted that if the lab fell inactive for more than two years, the collection would revert to the museum. All parties cheerfully accepted these terms, and final transfer of the specimens was made in July.[25]

Other collections also came to the Tree-Ring Lab. While arranging for Colton's collection, Douglass succeeded in obtaining specimens gathered by W.S. Stallings and deposited with the Laboratory of Anthropology in Santa Fe. By the end of 1950, over 9000 new specimens had been added to the Tree-Ring Lab's collection, and Terah L. Smiley, who had coordinated much of the transfer, was appointed Curator of Archaeological Collections. Harold S. Gladwin's collection from Gila Pueblo and his specimens on loan to Cal Tech also passed to the University of Arizona by 1957.[26]

Douglass urged further development of the Tree-Ring Lab. Although the lab budget more than doubled between 1950 and 1957, funds for capital improvements, travel and other expenses remained fairly constant during the period. Salaries accounted for the bulk of budgetary requests. In addition to Douglass, Schulman, Smiley and a half-time secretary, the staff grew to include Bryant Bannister as an archaeological assistant in 1954 and Marvin A. Stokes as Research Assistant in Prehistoric Dating two years later. Salaries for all staff members grew as quickly as the university would allow.[27] The Tree-Ring Lab appeared to be in excellent condition.

The institutional growth of dendrochronology was boosted during this period through the work of Douglass's chief associate, Edmund Schulman. While Douglass had turned his interest to the study of astronomic and climatological data,

Schulman had continued the purely dendrochronological work of the Tree-Ring Lab. Returning from a collection trip to Sun Valley, Idaho, in the summer of 1953, Schulman visited the White Mountains of east-central California. Here he found several bristlecone pines which dated at 1500 years of age. For the next two years, Schulman analyzed these and other specimens, and concluded that the trees provided very sensitive records of climatic change.[28]

By early 1956, Schulman's investigations had proved sufficiently intriguing that the National Science Foundation approved a three-year grant of $18,800 for his "Millennium-Long Tree-Ring Histories of Climatic Changes." That summer, he returned to the Inyo National Forest, convinced that he would find 4000-year-old trees. Unfortunately, the summer's collection only added 200 years to the previously determined record, extending it to approximately A.D. 50. The next summer's field season, however, proved far more productive. In the driest part of the forest, along a path called Methuselah Walk, Schulman discovered a 4000-year-old specimen with its center intact. Later analysis showed the tree to be over 4600 years old, the oldest living thing then discovered. By the end of the year, when Schulman had completed an article for *National Geographic*, seventeen trees had been found over 4000 years old.[29]

Schulman's success with the bristlecone pine, although symbolizing the significant position gained by the Laboratory of Tree-Ring Research, proved a bittersweet victory. On January 8, 1958, a few weeks after completing his article, Schulman suffered a fatal heart attack on the university campus. His associate's death heightened Douglass's concern for the future of the Tree-Ring Lab. Well into his ninety-first year, he realized his active oversight of the lab could not long continue. In late February, Douglass expressed his wish to retire from the directorship after June 30, leading President Richard A. Harvill to appoint Terah Smiley as acting director.

Douglass spent the next few months fretting over the future direction of the Laboratory for Tree-Ring Research. He soon became convinced that the lab's most important work involved the continuing search for climatic cycles. Although impressed with Smiley's talent and archaeological achievements, Douglass believed that the only colleague who possessed sufficient knowledge of astroclimatology to continue this study was Bryant Bannister, currently completing his doctorate. Douglass urged the rapid promotion of his assistant to the eventual directorship of the lab, convinced that this represented the only course to guarantee the continuation of important climatic work. Bannister's lack of administrative experience and age delayed his appointment until 1964, after the directorships of Terah Smiley (1958–1960) and William G. McGinnies (1960–1964).[30]

Although Douglass continued to worry about the future of the Tree-Ring Lab after his retirement, he welcomed the extra time he could devote to research. He completed his manuscript on astroclimatology during the summer of 1958, devoting much of the next year to revisions. Entitled "Climatic Cycles in the Southwestern United States: Based on Observations of the Sun, Planets and Tree Rings," the treatise sought to present a concise theory of the various astronomical influences on weather which he had perceived over several decades. The unifying concept in Douglass's hypothesis was the effect of planetary combinations on the sun. Conjunctions or oppositions, he maintained, disturbed the solar surface, creating sunspots and thus affected terrestrial weather. The tree-ring record reflected this influence. Douglass regarded the discovery of this planetary-solar relationship as essential to his goal of long-range weather forecasting. The connection between sunspots and rainfall had been established to his satisfaction for many years; now Douglass proposed the missing piece of the climatic puzzle:

> The astronomical mechanism which does this [causes sunspots and hence rain] is simply the gravitational (and perhaps other) forces that the two larger planets, Jupiter and Saturn, exert on the sun's surface. This seems to me to be the chief mechanical operation that controls our natural water supply by an astronomical event.

Further review of his work, however, revealed that Douglass's theories were far too speculative. The manuscript remained unpublished.[31]

Throughout the completion and revision of his manuscript, Douglass had been the victim of failing health from extreme old age. He continued to explore the world and universe mentally, but he was forced to curtail the active life he had led for so many years. Remarkably, Douglass had suffered no major health problems, except the severe cataract disabilities in his left eye. Even after corrective surgery, his eyesight remained his most enervating infirmity. After his formal separation from the Tree-Ring Lab, Douglass increasingly spent more time working at home than in his laboratory, but he continued to investigate and reflect. In the fall of 1959, however, even this activity slowed to a stop. An unavoidable bladder operation proved a severe shock to the entire system of the 92-year-old astronomer, a shock from which he never fully recovered.[32]

For the next eighteen months, Douglass occasionally had flashes of energy and interest, but these merely emphasized the slow erosion of his intellectual and physical powers. By the late summer of 1961, his wife wrote to their valued friend Harold S. Colton: "He is in a pitifully tragic mental state now. . . . [H]e forgets, and talks incoherently." A month later, his condition worsened and his wife regretfully placed him in a Tucson rest home.[33]

A.E. Douglass clung to life for several months, but his death on March 20, 1962, surprised no one. Notices of his passing appeared in virtually every newspaper in Arizona, as well as major periodicals throughout the world. Long discussions of his career filled columns in the *New York Times* and *The Times* of London. After a long life in the service of science and humanity, Douglass was quietly laid to rest in a Tucson cemetery. Soon after, the Tree Ring Society passed a memorial resolution which expressed the feelings of the scientific world:

> . . . the Tree-Ring Society herewith expresses its sense of deep loss in the death of Dr. Andrew Ellicott Douglass who some 57 years ago initiated the study of dendrochronology and who throughout this period by his wisdom, energy, and dedication directed the development and destinies of this important branch of science, a study which gave unique meaning to archaeology and which has materially contributed to such diverse fields as astronomy, climatology, and ecology.[34]

For seven decades Andrew Ellicott Douglass had actively served the scientific profession as astronomer and dendrochronologist. His contributions to both disciplines firmly established his reputation as a significant figure in the development of science in the twentieth century. His accomplishments represented a fitting monument to a long career devoted to the study of terrestrial and cosmic phenomena. Other monuments included two major observatories and the Laboratory of Tree-Ring Research, the focal point of an international network of dendrochronological research. Few scientists establish enviable reputations in two fields; fewer still create an entirely new discipline. Douglass's success in achieving both these feats assures his place among the leaders of modern science.

Notes To The Chapters

1. Various material, Box 2, Andrew Ellicott Douglass Papers, Special Collections, University of Arizona Library UAL, Tucson. Bryant Bannister, "Andrew Ellicott Douglass (1867–1962)," *Year Book of the American Philosophical Society* (1965), 121; Catharine Van Cortland Mathews, *Andrew Ellicott: His Life and Letters* (New York: The Grafton Press, 1908), 31–104; William B. Shaw, "Douglass, David Bates," *Dictionary of American Biography* (21 vols., New York: Charles Scribner's Sons, 1973), V, 405–406.
2. William A. Ellis (ed.), *Norwich University, 1819–1911: Her History, Her Graduates, Her Roll of Honor* (3 vols., Montpelier, Vermont: privately printed, 1911), III, 50–51. *The Church Kalendar* (Westfield, New York), October 8, 1887, copy in folder 9, Box 6, Douglass Papers, UAL.
3. Various material in Boxes 2–3; "Personal Story for the Press," ms. dated October 1929, Box 155, Douglass Papers, UAL.
4. Trinity College entrance examinations, June 24, 1884, June 23, 1885, Box 5; Douglass to Malcolm Douglass, June 28, 1885, Box 6; "Diagram of Eclipse, through Small Telescope," March 16, 1885, Box 9, Douglass Papers, UAL.
5. Grade report, Trinity (spring) term, 1887; Lewis H. Paddock to Douglass, June 8, 1888; Commencement Program, June 27, 1889,

Box 5, Douglass Papers, UAL. *Catalogue of the Officers and Students of Trinity College 1889–90* (Hartford: Trinity College, 1889), 45–61, passim.

6. Malcolm Douglass to Douglass, November 9, 1885, Box 6; Seance at Mrs. Helen Fairchild's," July 24, 1885; "Mesmerism," notes dated 1885–86, Box 9; Notes on Seybert Commission on Spiritualism, 1887, Box 8, Douglass Papers, UAL.
7. Douglass to Reverend and Mrs. Malcolm Douglass, October 4, 1885; Malcolm Douglass to Douglass, November 18, 1885, Box 6; Douglass to Martin Taylor, October 11, 1950, Box 7, Douglass Papers, UAL.
8. Arthur Berry, *A Short History of Astronomy: From Earliest Times Through the Nineteenth Century* (New York: Dover Publications, Inc., 1961; orig. pub. 1898), 371–72, 376–80, 402; Charles A. Whitney, *The Discovery of Our Galaxy* (New York: Alfred A. Knopf, 1971), 179–90.
9. Malcolm Douglass to Douglass, January 25, February 11, 1886, Box 6; "Mars. Old Observations," Spring 1886; "Observations on Sun," June 1886, Box 9, Douglass Papers, UAL.
10. Various photographs in folder 5, Box 5; "Personal Story for the Press," Box 155; F. S. Luther to Sarah Hale Douglass, December 17, 1891, Box 7, Douglass Papers, UAL.
11. Sarah Hale Douglass to Douglass, February 6, 1888, June 6, 1889, Box 6; Prosser H. Frye to Douglass, August 8, 1889, Box 7, Douglass Papers, UAL.
12. "Personal Story for the Press," Box 155; William H. Pickering to Douglass, September 17, 1889, Box 14, Douglass Papers, UAL. Douglass to Edward C. Pickering, July 7, September 15, 1889, August 22, 1890, E. C. Pickering Papers, Observatory Archives, Harvard University Archives [HUA]. Berry, *Short History of Astronomy,* 402–403.
13. Solon I. Bailey, *The History and Work of Harvard Observatory, 1839 to 1927* (New York: McGraw-Hill Book Company, Inc., 1931), 55, 281–82; William H. Pickering, "The Mountain Station of the Harvard College Observatory," *Astronomy and Astro-Physics [AAP],* V (May 1892), 353–54.
14. Pickering, "The Mountain Station," *AAP,* V, 353.
15. Diary, Book I (December 1890–May 1891); Douglass to Sarah Hale Douglass, January 4, 18, 1891, Box 10, Douglass Papers, UAL.

16. Douglass to Mary Carolyn Douglass, February 2, 1891; to Sarah Hale Douglass, February 28, May 9, August 31, 1891; to David Bates Douglass, March 24, 1891, Box 10, Douglass Papers, UAL. Bailey, 60.
17. Bailey, 36, 42, 58, 154–55, 158, 207–11; Whitney, 157–60, 165–76.
18. Diary, Book II (February–March 1891); Douglass to Sarah Hale Douglass, January 23, August 31, 1891, Box 10, Douglass Papers, UAL. Bailey, 61, 90; William H. Pickering, "The Planets," *Annals of the Astronomical Observatory of Harvard College [AAOHC]*, XXXII, part II (1900), 159–60; William K. Hartmann, *Moons and Planets: An Introduction to Planetary Science* (Belmont, California: Wadsworth Publishing Company, Inc., 1972), 10–13.
19. Various notes, folder 9, Box 12, Douglass Papers, UAL. William H. Pickering, "Brighter Satellites of Jupiter and Saturn," *AAOHC*, LXI, part I (1908), 75–81; Pickering, "The Planets," 169–70; Hartmann, 18, 23, 396–97; Bailey, 90.
20. William Graves Hoyt, *Lowell and Mars* (Tucson: The University of Arizona Press, 1976), 5–10; "Schiaparelli's Observations of Mars," *The Observatory, A Monthly Review of Astronomy*, V (May 1, 1882), 138–43. For an overview of Douglass's Martian research, see G. E. Webb, "A. E. Douglass and the Canals of Mars," *The Astronomy Quarterly*, III, number 9 (1979), 27–37.
21. Various notes, folder 9, Box 12, Douglass Papers, UAL. W. H. Pickering, "The Arequipa Observations of *Mars* and *Jupiter*," Astronomical Society of the Pacific, *Publications* [ASP *Publications*], VI (August 15, 1894), 222; Edward S. Holden, "Addendum—The Arequipa Observations of *Mars* and *Jupiter* in 1892," ibid., 226; W. H. Pickering, "Artificial Disks," *AAOHC*; XXXII, part II (1900), 121–24.
22. Diary, Book I; Douglass to Sarah Hale Douglass, August 31, 1891; to Mary Carolyn Douglass, February 11, 1891, Box 10, Douglass Papers, UAL.
23. Douglass to Sarah Hale Douglass, April 1, 21, 1891, March 9, 1892; Diary, Books 3 (April–July 1891), 5 (January–May 1892), 6 (June–October 1892), 7 (October–November 1892); Douglass to Benjamin Hale Douglass, April 29, 1891, Box 10; F. W. Putnam to Douglass, March 27, 1891, Box 12, Douglass Papers, UAL; Douglass to Edward C. Pickering, December 21, 1892, Pickering Papers, HUA.
24. Sarah Hale Douglass to Douglass, October 8, 1892; February 15, 1893, Box 10, Douglass Papers, UAL.

25. Diary (1893), Box 10, Douglass Papers, UAL; Douglass to E.C. Pickering, July 19, August 11, December 22, 1893, Pickering Papers, HUA.
26. A. Lawrence Lowell, *Biography of Percival Lowell* (New York: The Macmillan Company, 1935), 1–60; Hoyt, 15–17.
27. George T. Vickers to Douglass, February 19, 1894, Box 11; "Reminiscences of the Lowell Observatory," ms. dated 1920 in Box 17; "Arizona in the Nineties and the Beginning of the Lowell Observatory," ms. dated November 12, 1947, folder 1, Box 22, Douglass Papers, UAL. Douglass to E.C. Pickering, December 22, 1893, March 14, 1894, Pickering Papers, HUA. Hoyt, 26.
28. Lowell, *Percival Lowell*, 62–64, 70–71; Hoyt, 39–46; A.E. Douglass "The Study of Atmospheric Currents by the Aid of Large Telescopes, and the Effect of Such Currents on the Quality of Seeing," *The American Meteorological Journal*, XI (March 1895), 395–96.
29. A.E. Douglass, "Gegenschein Observations," *Popular Astronomy* [PA], II (September 1894), 29–32; Notebook No. 1, A.E. Douglass Papers, Lowell Observatory Archives [LOA], Flagstaff, Arizona.
30. A.E. Douglass, "The Lowell Observatory and Its Work," *PA*, II (May 1895), 402; Hoyt, 316; Notebook No. 1, Douglass Papers, LOA.
31. Notebook No. 1, Douglass Papers, LOA. Douglass to Percival Lowell, March 9, 1894, Percival Lowell Papers, LOA.
32. Douglass to Lowell, March 9, 20, 23, 1894, Lowell Papers, LOA.
33. Douglass to Lowell, March 9, 10, 1894, Lowell Papers, LOA.
34. Notebook No. 1, Douglass Papers, LOA; Douglass to Lowell, March 12, 1894, Lowell Papers, LOA; Tucson *Arizona Weekly Citizen*, March 10, 1894; Tuscon *Arizona Citizen*, March 13, 1894.
35. Notebook No. 1, Douglass Papers, LOA; Douglass to Lowell, March 14–16, 1894, Lowell Papers, LOA; Tucson *Arizona Citizen*, March 14–17, 1894; Tucson *Arizona Weekly Citizen*, March 17, 1894.
36. Douglass to Lowell, March 13, April 9, 1894, Lowell Papers, LOA; Theodore B. Comstock to Douglass, March 27, 1894; George J. Roskruge to Douglass, April 4, 1894, Douglass Papers, LOA; Tucson *Arizona Weekly Citizen*, March 17, 31, April 7, 14, 1894.
37. Notebook No. 1; W.J. Van Horn to Douglass, February 20, 1894, Douglass Papers, LOA; Phoenix *Arizona Republican*, March 18, 1894.

38. Douglass to Lowell, March 19, 23, 1894; Lowell to Douglass, March 20, 1894, Lowell Papers, LOA.
39. Notebook No. 1, Douglass Papers, LOA; Douglass to Lowell, March 23, 26, 1894; Lowell to Douglass, March 24, May 24, 1894, Lowell Papers, LOA.
40. Notebook No. 2, Douglass Papers, LOA; "Lowell Observatory (Remeniscences [*sic*])," ms. dated 1944, Box 17, Douglass Papers, UAL; Prescott *Arizona Journal-Miner,* March 28, 29, 1894.
41. Notebook No. 2, Douglass Papers, LOA. Douglass to Lowell, March 30, April 1, 1894; Lowell to Douglass, March 28, 1894, Lowell Papers, LOA. Prescott *Arizona Journal-Miner,* March 31, April 2, 1894.
42. Notebook No. 2, Douglass Papers, LOA; Lowell to Douglass, April 6, 7, 1894; Douglass to Lowell, April 6, 1894, Lowell Papers, LOA.
43. Douglass to Lowell, April 16, 20, 1894; Lowell to Douglass, April 16, 21, 1894, Lowell Papers, LOA; *Annals of the Lowell Observatory* [*Annals*], I (Boston: Houghton, Mifflin and Company, 1898), viii.

Chapter Two

1. "Reminiscences of the Lowell Observatory," (1920), Box 17, Douglass Papers, UAL; *Annals,* viii; Phoenix *Arizona Gazette,* March 22, 1894.
2. *Annals,* I, vii–xi.
3. Notebook No. 2, Douglass Papers, LOA; *Annals,* I, viii.
4. Douglass to Lowell, May 16, 17, 19, 1894, Lowell Papers, LOA; *Annals,* I, viii; Flagstaff *Coconino Weekly Sun,* May 17, 1894.
5. Flagstaff *Coconino Weekly Sun,* April 26, 1894; T.R. Gabel to Douglass, May 3, 1894, Douglass Papers, LOA; Douglass to Lowell, May 10, 1894; Lowell to Douglass, May 17, 18, 1894, Lowell Papers, LOA.
6. Douglass, "The Lowell Observatory and Its Work," *PA,* II, 395–98; *Annals,* I, ix; Flagstaff *Coconino Weekly Sun,* May 24, 31, 1894.
7. *Annals,* I, 3–4; Hoyt, *Lowell and Mars,* 61.
8. *Annals,* I, 9, 45, 80–86, 95, 100, 208–13, 247-50; Lowell to Douglass March 23, 1895, Lowell Papers, LOA.

9. *Annals*, I, 59–66, 74–75; Percival Lowell, "On the Existence of a Twilight Arc upon the Planet Mars," *Astro-Physical Journal* [*AJ*], II (August 1895), 136, 143-45; Robert S. Ball, "Mars," ASP *Publications*, V (January 28, 1893), 34.
10. *Annals*, I, 200, 213–15, 253, 256–58, 273; J.M. Schaeberle, "Preliminary Note on the Observations of the Surface Features of *Mars* during the Opposition of 1892," ASP *Publications*, IV (September 3, 1892), 197.
11. *Annals*, I, 203; Douglass, "Lowell Observatory," *PA*, III, 401. Douglass to Lowell, December 24, 1894; Lowell to Douglass, January 12, 1895, Lowell Papers, LOA.
12. *Annals*, I, 199–202.
13. *Annals*, 293–94, 337–38; Lowell, "Twilight Arc," *AJ*, 136–47; Douglass, "Lowell Observatory," *PA*, II, 400; Douglass to Lowell, January 23, August 23, 1895, Lowell Papers, LOA.
14. *Annals*, I, 338–39, 349–56, 362–74; A.E. Douglass, "A Cloud-Like Spot on the Terminator of Mars," *AJ*, I (February 1895), 127–30.
15. Henry H. Bates, "The Chemical Constitution of Mars' Atmosphere," ASP *Publications*, VI (December 1, 1894), 301.
16. Edward S. Holden, "The Lowell Observatory in Arizona," ASP *Publications*, VI (June 9, 1894), 164; W.W. Campbell, "The Spectrum of Mars," ibid., VI (August 15, 1894), 236; Campbell, "Concerning an Atmosphere on Mars," ibid., VI (December 1, 1894), 273–83.
17. George Abell, *Exploration of the Universe* (Brief edition, New York: Holt, Rinehart and Winston, 1969), 214–15; *Annals of the Lowell Observatory*, II (Cambridge: The University Press, 1900), 4, 23, 157–58.
18. *Annals*, II, 10–13, 27–28, 69–83, 101–111, 127–32, 194, 200; Douglass, "Lowell Observatory," *PA*, II, 401. Bradford A. Smith, et al., "The Galilean Satellites and Jupiter: Voyager 2 Imaging Science Results," *Science*, 206 (November 23, 1979), 927–50.
19. A.E. Douglass, "A Combination Telescope and Dome," *AJ*, I (May 1895), 401–10. Douglass to Lowell, January 14, February 2, 20, 1895; Lowell to Douglass, February 8, 1895, Lowell Papers, LOA.
20. Flagstaff *Coconino Weekly Sun*, May 17, June 21, 28, August 2, 9, 16, 23, 30, September 6, 20, October 4, 18, November 8, December 20, 1894; March 1, 1895.
21. Lowell to Douglass, January 5, 1895; Douglass to Lowell, September 6, December 22–31, 1894, January 4, 8, 11, February 5, 9, 22,

March 13, 24, April 3, 1895, Lowell Papers, LOA.

22. Douglass to Lowell, January 9, 17, 23, 29, February 12, 1895, Lowell Papers, LOA.
23. Douglass to Lowell, December 30, 1894, Lowell Papers, LOA.
24. Lowell to Douglass, March 5, 21, 1895, Lowell Papers, LOA.
25. Flagstaff *Coconino Weekly Sun,* March 28, April 4, 1895. Douglass to Lowell, April 12, 13, 15, May 4, 8, 1895; Lowell to Douglass, May 4, 7, 1895 (telegrams), Lowell Papers, LOA.
26. Douglass to Lowell, March 25–27, 1895, Lowell Papers, LOA.
27. Douglass to Lowell, March 28, 1895, Lowell Papers, LOA.
28. Douglass to Lowell, April 21, 22, 29, May [?(telegram)], 2, 1895; Lowell to Douglass, May 1, 4, 1895 (telegrams), Lowell Papers, LOA.
29. Douglass to Lowell, May 16, 28, June 3, 12, 1895, Lowell Papers, LOA. Notebooks No. 3, 4, Douglass Papers, LOA.
30. Lowell to Douglass, March 4 (telegram), June 1, July 2 (telegram), 1896, Lowell Papers, LOA. A. Lawrence Lowell to Douglass, March 9, 1896 (telegram), Douglass Papers, LOA.
31. Flagstaff *Coconino Weekly Sun,* July 16, 23, October 15, November 16, 1896; T.J.J. See, "A Sketch of the New 24-inch Refractor of the Lowell Observatory," *PA,* IV (December 1896), 297–300; Hoyt, *Lowell and Mars,* 103–105; *Annals,* II, 207.
32. Felipe Valle to Douglass, June 16, 1896, Douglass Papers, LOA. Lowell to Douglass, November 19 (telegram), 26, 1896, Lowell Papers, LOA. A.E. Douglass, "The Lowell Observatory in Mexico," *PA,* IV (March 1897), 489.
33. Douglass, "Lowell Observatory in Mexico," *PA,* IV, 490–94; Flagstaff *Coconino Weekly Sun,* November 12, 19, 1896.
34. Douglass, "Lowell Observatory in Mexico," *PA,* IV, 490–91. Don A. Sweet to Douglass, November 4, 1896 (telegram), Douglass Papers, LOA.
35. Douglass, "Lowell Observatory in Mexico," *PA,* IV, 490–92. Lowell to Douglass, December 12, 1896, Lowell Papers, LOA.
36. Douglass, "Projections on the Terminator of Mars and Martian Meteorology," *Astronomische Nachrichten* [*AN*], CXLII (1897), No. 3406, 363–66; *Annals,* II, 236–37, 441, 447, 462–66, 499–507, 511.
37. Douglass, "Planetary Work at Lowell Observatory," *PA,* VII, 78–80; Douglass, "The First Satellite of Jupiter," *AN,* CXLVI (1898), 346–56; Douglass, "Drawings of Jupiter's Third Satellite,"

AN, CXLIII (1897), 411–14. W.H. Pickering to Douglass, December 31, 1897, Box 14; "Unpublished Work," mss. in Box 17, Douglass Papers, UAL.

38. Douglass, "An Ascent of Popocatapetl," *PA*, V (February 1898), 505-508; Douglass, "Effects of High-Mountain Climbing," *Appalachia*, VIII (March 1898), 3; Douglass, "The Altitudes of Orizaba and Popocatapetl," ibid., 356–61; W.A. Cogshall, "A Trip to the Summit of Orizaba," ibid., 350–55. Lowell to Douglass, April 21, 1897 (telegram), Lowell Papers, LOA.

Chapter Three

1. *Flagstaff Sun-Democrat*, March 18, April 22, 1897. Hoyt, *Lowell and Mars*, 114.
2. *Flagstaff Sun-Democrat*, May 6, 13, July 1, 1897. Douglass to Lowell, May 17, 1897 (telegram); Lowell to Douglass, July 2, 1897, Lowell Papers, LOA.
3. *Flagstaff Sun-Democrat*, November 18, 1897. *San Diegan-Sun* (California), November 20, 1897, in scrapbook, Box 21a, Douglass Papers, UAL. Lowell to Douglass, August 9, 1897, Lowell Papers, LOA. W.L. Putnam to Douglass, November 15 (telegram), December 5, 1897, Letter Book #3, LOA.
4. Hoyt, *Lowell and Mars*, 122. Putnam to Douglass, June 14, 1898, Douglass Papers, LOA. Putnam to Douglass, July 8, 1898, Letter Book #3, LOA.
5. Douglass to Putnam, June 28, 1898; to Henry S. Pritchett, August 3, 1898, Douglass Papers, LOA. Putnam to Douglass, July 8, 1898, Letter Book #3, LOA. John Lankford, "A Note on T.J.J. See's Observations of Craters on Mercury," *Journal for the History of Astronomy*, XI (June 1980), 129.
6. Lowell to Douglass, April 21, May 28, June 1, July 19, 1897, Lowell Papers, LOA. *Annals of the Lowell Observatory*, III (privately printed, 1905), "Supplement," 1, 24–25; A.E. Douglass, "Present Rotation-Period of the First Satellite of Jupiter and its Change in Form and Period Since 1892," *Astronomical Journal*, XIX (August 1, 1898), 69–70.
7. Flagstaff *Coconino Sun*, December 31, 1898; *Boston Evening Transcript*, January 27, 1900. W.R. Warner to Douglass, March 21, 1899; Godfrey Sykes to Douglass, April 4, May 11, 1899; S.L. Boothroyd to Douglass, June 2, 1899, Douglass Papers, LOA.

8. Lowell to Douglass, June 1, 1897, Lowell Papers, LOA. Notebook No. 4, Douglass Papers, LOA.
9. Flagstaff *Coconino Sun*, June 4, October 1, 1898; September 1, 1900; May 18, 1901.
10. Putnam to Douglass, January 19, April 13, May 26, July 8, August 1 (telegram), 1898, Letter Book #3, LOA. W. Louise Leonard to Douglass, May 25, 1899, Douglass Papers, LOA.
11. *Annals*, II, x. Putnam to Douglass, October 2, 12, 1897; January 26, 1898; June 16, 1899, Letter Book #3, LOA. W. Louise Leonard to Douglass, May 31, June 2, Letter Book #4, LOA.
12. Putnam to Douglass, June 14, 1898, Putnam Correspondence, LOA. Winslow Upton to Douglass, September 6, 1898; William W. Payne to Douglass, August 8, 1898; March 1, 1899, Douglass Papers, LOA.
13. Douglass, "The Astronomer's Globe," *PA*, V (June 1897), 57–62; Douglass, "Atmosphere, Telescope and Observer," ibid., 64–84.
14. Douglass, "Mars," *PA*, VII (March 1899), 113–17; Douglass, "An Hypothesis Regarding the Surface Markings of Jupiter," ibid., VIII (November 1900), 473–75; Bradford A. Smith, et al., "The Jupiter System Through the Eyes of Voyager I," *Science*, 204 (June 1, 1979), 951–972.
15. Douglass, "Planetary Work at Lowell Observatory," *PA*, VIII, 83–85; Douglass, "The Effect of Mountains on the Quality of the Atmosphere," *PA*, VII (August 1899), 355–56, 363–64.
16. Douglass, "The Lick Review of 'Mars,'" *PA*, IV (October 1896), 199–201. Draft of "The Lick Review of 'Mars,'" Miscellaneous Papers, LOA.
17. Douglass, "Atmosphere, Telescope and Observer," *PA*, V, 65; Douglass, "Mars," ibid., VII, 113.
18. Douglass, "Planetary Work at Lowell Observatory," *PA*, VII, 80–82; Douglass, "The Markings on Venus," *Monthly Notices of the Royal Astronomical Society*, XVIII (May 13, 1898), 382–85.
19. Instructions for summer, 1900, Box 16; "Unpublished Work," mss., Box 17; Notes on Saturn, July–August, 1900, Box 19, Douglass Papers, UAL.
20. Douglass to Lowell, December 8, 1900 (telegram), Lowell Papers, LOA. *Annals*, III, 26, 101, 104–105, 143; Percival Lowell, "Explanation of the Supposed Signals from Mars of December 7 and 8, 1900," *American Philosophical Society [APS] Proceedings*, XL (December 1901), 166–76.

21. Daniel E. Parks to Douglass, January 17, 1901, Douglass Papers, LOA. Various newspaper clippings in Box 16, Douglass Papers, UAL. *New York Times*, January 16, 1901.
22. Douglass to Putnam, March 12, 1901, Box 16, Douglass Papers, UAL. W.W. Campbell, "*Mars*, By Percival Lowell," ASP *Publications, VIII* (August 1, 1896), 208–209, 215, 217–19.
23. E.E. Barnard, "On the Third and Fourth Satellites of Jupiter," *AN*, CXLIV (1897), 322; William H. Pickering, "Artificial Disks," *AAOHC*, XXXII, Part II (1900), 150–55.
24. Douglass to Lowell, December 31, 1894; Lowell to Douglass, January 5, 1895, Lowell Papers, LOA. *Annals*, I, 6, 79, 193, 196; *Annals*, III, 22–23.
25. Douglass to Lowell, October 16, 1900; Memorandum dated 1/7/12, Box 16; "Unpublished Work," mss. dated 1906, Box 17; "Artificial Planets," Box 19, Douglass Papers, UAL.
26. Hoyt, *Lowell and Mars*, 51, 123–24. Douglass to Joseph Jastrow, January 9, 1901, Douglass Papers, UAL.
27. Douglass to W.H. Pickering, March 8, 1901; Pickering to Douglass, March 27, 1901, Box 14, Douglass Papers, UAL.
28. Douglass to Putnam, March 12, 1901, Box 16, Douglass Papers, UAL.
29. "Lowell Obsy Matter," mss. dated 1/1/29; Douglass to Lowell, September 21, 1901, Box 16, Douglass Papers, UAL. W.L. Leonard to Douglass, September 8, 1901, Letter Book #4, LOA. Flagstaff *Coconino Sun*, August 10, 1901. On the reality of the Martian Canals, see Webb, "A.E. Douglass and the Canals of Mars," 34–35.
30. Douglass to W.W. Campbell, August 23, 1901, Mary Lea Shane Archives of the Lick Observatory, University of California, Santa Cruz. Douglass to E.C. Pickering, December 4, 1901 (telegram), Pickering Papers, HUA. Flagstaff *Coconino Sun*, December 14, 1901.
31. Flagstaff *Coconino Sun*, November 9, 1901; February 1, June 28, 1902. Various material in Box 24, Douglass Papers, UAL.
32. Flagstaff *Coconino Sun*, September 27, 1902.
33. Ibid., October 11, November 1, 8, 22, 1902.
34. Ibid., September 10, November 12, 26, 1904. Various material, folder 6, Box 22, Douglass Papers, UAL.
35. Douglass to E.C. Pickering, January 5, March 26, April 28, June 16, 1903; Pickering to Douglass, January 10, 1903, Pickering Papers, HUA.

36. R.H. Tucker to Douglass, July 24, 1905, Lick Observatory Archives. Douglass to U.S. Hall, August 28, 1905, Box 23, Douglass Papers, UAL. Lowell to Douglass, November 7, 1905, Letter Book #7, LOA.
37. Flagstaff *Coconino Sun*, July 5, 1902; August 1, 1903; July 16, 1904; July 8, August 5, 12, 1905. Various material in folder 1, Box 175; material on Joseph C. Whittington, folder 4, Box 150, Douglass Papers, UAL.
38. Douglass to Edwin L. Lord, March 19, 1905, Box 24; T.J. Coalter to Douglass, May 13, 1905, Box 23; various material in folder 11, Box 22, Douglass Papers, UAL.
39. Douglass to E.S. Clark, August 19, 24, 1905; Clark to Douglass, August 23, 26, 1905; Henry Fountain Ashurst to Douglass, September 9, 1905, Box 23, Douglass Papers, UAL. *Annual Announcement, Northern Arizona Normal School, 1905–1906* (Flagstaff: C.M. Funston, Printer, 1905), 9, 18–19.
40. Douglass to Orville Brewer, February 2, 1906; William F. Peirce to Douglass, May 14, 1906, Box 23; Luther to Douglass, May 24, 1906, Box 7, Douglass Papers, UAL. Flagstaff *Coconino Sun*, May 12, 26, 1906.

Chapter Four

1. "Atmospheric Currents 1906–1907, 1915," folder 1, Box 50, Douglass Papers, UAL.
2. *Tucson Citizen*, February 14, 1907. Douglass to Committee on Appropriations of the Legislature, 1907, Box 44, Douglass Papers, UAL.
3. Tucson *Arizona Star*, April 12, 1908.
4. George W.P. Hunt to Douglass, December 17, 1908; Douglass to Hunt, December 31, 1908, Box 152, Douglass Papers, UAL.
5. Douglass to Mrs. Russell Sage, February 12, 1910; to Merrill P. Freeman [1911?], Box 44, Douglass Papers, UAL. Tucson *Arizona Star*, February 6, 1910; A.E. Douglass, "Historical Address," *The Inaugural Bulletin* (Tucson: University of Arizona, 1923), 54.
6. Douglass to A.H. Wilde, December 9, 1911, Box 27, Douglass Papers, UAL.
7. Wilde to Douglass, December 13, 1911, Box 27, Douglass Papers, UAL.

8. W.H. Pickering to Douglass, March 11, 1908; Douglass to E.C. Pickering, May 13, 1908, Box 44, Douglass Papers, UAL. Tucson *Arizona Star*, May 10, 1908.
9. Douglass, "Historical Address," 54; Williard P. Gerrish to Douglass, November 19, December 5, 1908; January 9, March 9, 1909, Box 44, Douglass Papers, UAL.
10. Douglass, "Historical Address," 54–55. Gerrish to Douglass, April 9, 1909; "Sketches: Driving Clock for the 8-inch;" "The Harvard 8" Telescope," ms. dated May 12, 1945, Box 44; various material in folder 8, Box 52, Douglass Papers, UAL.
11. "Record of Seeing," Box 51; "The Division in the Nucleus of Halley's Comet and Other Observations," folder 3, Box 50, Douglass Papers, UAL. *Tucson Citizen*, May 18, 1910; Douglass, "Drawings of Comet *a*, 1910," *PA*, XVIII (March 1910), 162–63.
12. Various letters in folder 8, Box 50, Douglass Papers, UAL. *Tucson Citizen*, November 9, 1914.
13. Douglass to C.E.K. Mees, June 25, July 9, 1915; Mees to Douglass, July 1, 1915; C.A. Chant to Douglass, March 11, 1916, Box 54, Douglass Papers, UAL. Douglass, "Zodiacal Light and Counterglow and the Photography of Large Areas and Faint Contrast," *Photographic Journal*, LVI (February 1916), 44–48.
14. Douglass to J.A. Brashear, June 15, 1914; James B. McDowell to Douglass, July 1, 1914; "Plans 1914," Box 44, Douglass Papers, UAL.
15. Douglass to E.E. Ellinwood, June 5, 1914; Douglass to John T. Hughes, June 15, 1914; Hughes to Douglass, June 19, 1914, Box 44, Douglass Papers, UAL.
16. Douglass to Freeman, July 6, 1914, Box 44, Douglass Papers, UAL.
17. Douglass to George E. Hale, July 15, 1914; to McDowell, July 15, 1914; to E.B. Frost, July 15, 1914; to Campbell, July 15, 1914; W.H. Pickering to Douglass, July 25, 1914, Box 44, Douglass Papers, UAL.
18. Douglass to Hunt, July 25, August 3, 1914; Hunt to Douglass, August 4, 1914; Douglass to Philip Fox, August 20, 1914; copy of resolution by American Astronomical Society, August 28, 1914, Box 44, Douglass Papers, UAL.
19. Freeman to Douglass, October 17, 1914, Box 44, Douglass Papers, UAL.

20. Unidentified newspaper clipping, January 30, 1915, Box 44, Douglass Papers, UAL.
21. Douglass to Fox, October 17, 1916, Box 44, Douglass Papers, UAL. Tucson *Arizona Star*, October 19, 1916.
22. Douglass, "The Steward Observatory of The University of Arizona," ASP *Publications*, XXX (October 1918), 326–30; Tucson *Arizona Star*, August 12, 1917.
23. "Report to President—November 10, 1916," folder 6; "Plans for Immediate Operation of the Steward Observatory," folder 7, Box 44, Douglass Papers, UAL.
24. "Report to President—November 10, 1916," "Observatory Site Work 1916," folder 6, Box 44, Douglass Papers, UAL. *Tucson Citizen*, October 28, 1916; Douglass, "The Steward Observatory of The University of Arizona," ASP *Publications*, XXX, 327–28.
25. Hale to Douglass, November 3, 1916; George W. Ritchey to Douglass, November 6, 28, 1916, Box 44, Douglass Papers, UAL.
26. Douglass to J.A. Brashear Co., November 6, 16, 1916; McDowell to Douglass, October 24, November 11, 22, 1916; Douglass to McDowell, December 2, 1916, Box 44, Douglass Papers, UAL.
27. Douglass to Warner and Swasey, October 18, November 8 (telegram), 10 (telegram), 1916; L.B. Stauffer to Douglass, November 4, 11, December 30, 1916; Warner and Swasey to Douglass, November 9, December 13, 1916 (telegrams), Box 44, Douglass Papers, UAL.
28. Douglass to Hale, February 20 (telegram), March 5, 1917; "Plans and Notes 1917," Box 44, Douglass Papers, UAL. Douglass, "The Steward Observatory of The University of Arizona," ASP *Publications*, XXX, 329.
29. McDowell to Douglass, February 2, March 12, 1917; Douglass to John A. Brashear Co., March 5, 1917; J.F. Wrenshall to McDowell, April 2, 1917, Box 44, Douglass Papers, UAL.
30. Douglass to Corning Glass Works, July 17, 1917 (telegram); A.J. Mayer to Douglass, July 18, 1917; Douglass to von KleinSmid, July 26, 1917; von KleinSmid to Douglass, August 7, 1917, Box 44, Douglass Papers, UAL.
31. Douglass to Warner and Swasey, February 4, 27, March 3 (telegrams), 5, 19 (telegram); Stauffer to Douglass, February 5, 15, March 1, 1917; "Plans and Notes 1917," folder 9; Contract with Warner and Swasey, dated March 19, 1917, folder 10, Box 44, Douglass Papers, UAL.

32. Douglass to E.P. Burrell, June 23, 1917; to von KleinSmid, June 27, 1917, Box 44, Douglass Papers, UAL.
33. Stauffer to Douglass, July 31, October 26, 1917; Douglass to Warner, October 17, 1917, Box 44, Douglass Papers, UAL. Tucson *Citizen*, December 15, 1917.
34. Douglass to McDowell, March 1, 1918; McDowell to Douglass, March 12, 1918; Douglass to Warner, March 1, 1918; Warner to Douglass, March 13, 1918, Box 44, Douglass Papers, UAL.
35. Douglass to Warner and Swasey, November 11, 1918 (telegram); Stauffer to Douglass, November 12, 1918; Douglass to McDowell, November 18, 1918; McDowell to Douglass, December 19, 1918, Box 44, Douglass Papers, UAL.
36. Stauffer to Douglass, February 14, 1919; Warner and Swasey to Douglass, April 4, 1919 (telegram); Douglass to Warner and Swasey, April 8, 1919 (telegram); "Plans and Notes 1919," folder 2, Box 45, Douglass Papers, UAL.
37. Warner Seely to Douglass, July 15, 1919; P.E. Bliss to Douglass, September 16, 1919; Douglass to von KleinSmid, September 19, 1919; to Bliss, September 27, 1919; H.P. Bailey to Douglass, November 29, 1919, Box 45, Douglass Papers, UAL.
38. McDowell to Douglass, November 24, December 18, 1919; Douglass to McDowell, December 1, 1919; to Seely, December 1, 1919; Seely to Douglass, December 15, 1919, Box 45, Douglass Papers, UAL.
39. Douglass to von KleinSmid, April 18, 1919 (telegram); McDowell to Douglass, April 19, 1919, Box 45, Douglass Papers, UAL. Tucson *Citizen*, May 1, 1919.
40. McDowell to Douglass, May 29, July 21, 1919; Douglass to McDowell, June 24, 1919; George W. Morey to McDowell, July 17, 1919; Box 45, Douglass Papers, UAL.
41. McDowell to Douglass, September 11, October 20, December 29, 1919; H.N. Ott to Douglass, October 15, 1919; Douglass to Morey, December 3, 1919; St. Gobain, Chauny & Cirey to Brashear Co., June 12, 1919, Box 45, Douglass Papers, UAL.

Chapter Five

1. Richard E. Schmidt to Douglass, January 27, April 10, May 28, June 12 (telegram), 1920; Douglass to Schmidt, March 18, 31, April

20, May 25 (telegram), 1920; "Abstract" in "Correspondence relating to DOME CONSTRUCTION," Box 45, Douglass Papers, UAL.

2. Douglass to Seely, July 6, October 18, 1920; "Steinfeld Bills 1920," folder 4, Box 45, Douglass Papers, UAL. Douglass, "Historical Address," 57.
3. Douglass to Warner and Swasey, January 26, 29 (telegram), 1920; to Warner, November 27, 1920; to Seely, July 6, December 25 (abstract), 1920; Seely to Douglass, February 13, June 22, September 23, December 17, 28, 1920; "Summer 1920. Steward Observatory Trip," Box 45, Douglass Papers, UAL.
4. Douglass to H.N. Ott, January 12, 1920; Ott to Douglass, January 29, 1920; A.J. Mayer to John A. Brashear Co., Ltd., February 23, 1920; F.J.K. Bausch & Lomb to Douglass, March 13, 1920, Box 45, Douglass Papers, UAL.
5. Morey to Douglass, May 11, 1920; Douglass to Seely, May 29, 1920; to Ott, October 20, 1920; Ott to Douglass, October 26, 1920; McDowell to Douglass, December 2, 1920, Box 45, Douglass Papers, UAL.
6. Seely to Douglass, January 12, May 5, 1921; McDowell to Douglass, January 14, 22, 1921; Douglass to McDowell, January 17, 1921, Box 45, Douglass Papers, UAL.
7. Douglass to von KleinSmid, March 21, 1921; Seely to Douglass, June 1 (telegram), 1921; Douglass to Seely, May 19, 1921, Box 45, Douglass Papers, UAL. Tucson *Citizen*, January 22, 1921.
8. "Steward Observatory trip to inspect new telescope, Cleveland, Ohio;" Douglass to von KleinSmid, June 11, 1921 (telegram), Box 45, Douglass Papers, UAL.
9. Seely to Douglass, July 1, 11, 1921; Warner and Swasey to Douglass, July 15, 20, 26, 1921 (telegrams), Box 45, Douglass Papers, UAL.
10. Douglass to Warner and Swasey, July 25, 1921 (telegram); Warner and Swasey to Douglass, July 26, 1921; unidentified newspaper clipping, folder 7, Box 45, Douglass Papers, UAL. Tucson *Arizona Star*, July 23, 1921.
11. Tucson *Arizona Star*, July 23, 1921. Unidentified newspaper clipping, folder 7, Box 45, Douglass Papers, UAL.
12. "Notes on construction," dated August 1921, folder 7; "Clock Study Aug. 1921," folder 9; Warner and Swasey to Douglass, August 27, September 7, 8, 1921 (telegrams); Douglass to Warner

and Swasey, August 29, September 7, 1921 (telegrams), Box 45, Douglass Papers, UAL.

13. Ott to Douglass, January 4, February 12, 1921; Donald E. Sharp to Douglass, August 1, 18, October 3, 1921; McDowell to Douglass, April 11, May 24, 1921; Douglass to Ott, May 13, August 17 (telegram), 1921, Box 45, Douglass Papers, UAL.
14. Sharp to Douglass, October 3, November 17, December 19, 22 (telegram), 27, 1921; Sharp to McDowell, October 14, 1921, Box 45, Notes dated 1/24/22, folder 1, Box 46, Douglass Papers, UAL.
15. Sharp to Douglass, January 14, 30, 1922; McDowell to Douglass, February 21, April 25, May 12, June 6, 14, 28 (telegram), July 5, 1922; "Note of bill from Goodin Transfer Co.," July 10, 1922, folder 3, Box 46; "Report of Steward Observatory, 1922–23," folder 1, Box 47, Douglass Papers, UAL. Douglass, "Historical Address," 57–58.
16. Douglass "Steward Observatory," *UA Annual Report* (1939), 100. "Plans for the 36-inch," August 1923, folder 4, Box 46; "Report of Steward Observatory, 1922–23," July 3, 1923, Box 47, Douglass Papers, UAL.
17. Douglass to Hale, February 10, 1923, Box 13, George Ellery Hale Papers, California Institute of Technology Archives, Pasadena. Various material in folder 5; F. Louise Gianetti to Douglass, February 13, 1923; Campbell to Douglass, March 5, 1923; Douglass to W.S. Adams, March 9, 1923, Box 46, Douglass Papers, UAL.
18. "Invitation List," folder 5; "Draft of Dedication Program," Box 46, Douglass Papers, UAL. C.H. Marvin, "Address of Welcome," *Inaugural Bulletin*, 39; V.M. Slipher, "Response to Address of Welcome," ibid., 40.
19. R.G. Aitken, "Recent Progress in our Knowledge of the Universe," ibid., 42–53. Aitken's dedication address was also published in *Science*, November 16, 1923, 381–87.
20. Douglass, "Historical Address," 54–59.
21. "Report of Steward Observatory, 1922–23," July 3, 1923; "Report of Steward Observatory," June 4, 1924; "Steward Obsy Ann Report," June 3, 1925; "Steward Observatory and Department of Astronomy," March 27, 1927, Box 47, Douglass Papers, UAL.
22. Douglass to W.S. Eichelberger, September 20, 1918, Box 49, Douglass Papers, UAL. Douglass, "The University of Arizona

Eclipse Expedition. Port Libertad, Sonora, Mexico, September 10, 1923," ASP *Publications,* XXXVI (August 1924), 170.

23. Douglass to Joaquin Gallo, October 25, November 22, 1922; Notes dated May 2–7, 1923, Box 49, Douglass Papers, UAL. Douglass, "University of Arizona Eclipse Expedition," ASP *Publications,* XXXVI, 173.

24. Notes dated March 10, August 8, 1923, Box 49, Douglass Papers, UAL. Douglass, "University of Arizona Eclipse Expedition," ASP *Publications,* XXXVI, 170–71, 174–79; Glenton G. Sykes, "A Sonoran Adventure: The University of Arizona's Eclipse Expedition of September, 1923," *Mercury,* I (January/February 1972), 13–14.

25. "Notes from the Diary of Godfrey Sykes, with Added Remarks by Glenton G. Sykes, Member of the Expedition," undated ms., folder 5; Douglass to C.H. Marvin, September 7, 1923 (telegram), Box 49, Douglass Papers, UAL. Sykes, "A Sonoran Adventure," 15–16.

26. Douglass, "University of Arizona Eclipse Expedition," ASP *Publications,* XXXVI, 173, 179–81.

27. Ibid., 181–82. "Notes from the Diary of Godfrey Sykes," Box 49, Douglass Papers, UAL.

28. "University of Arizona Eclipse Expedition Expenses, 1923," Box 49, Douglass Papers, UAL. Douglass, "University of Arizona Eclipse Expedition," ASP *Publications,* XXXVI, 184.

29. "Report of Steward Observatory, 1923–24," June 4, 1924; "Steward Observatory Annual Report," June 3, 1925, Box 47; "Mars 1924 (Visual work—36-inch);" "Mars 1926," folder 5; "Mars (Photography of), 1924–1926," folder 9, Box 53, Douglass Papers, UAL.

30. Notes on Martian observations, 1931–39; E.P. Martz, Jr., to Douglass, May 21, 1939; February 6, 1940, Box 53, Douglass Papers, UAL. Philip C. Keenan, "The Measurement of Lunar Color Differences," ASP *Publications,* XLIII (June 1931), 203–14; Douglass, "Steward Observatory," 100–102.

31. "Steward Observatory Annual Report," June 3, 1925, Box 47, Douglass Papers, UAL. Berendzen and others, *Man Discovers the Galaxies,* 104–107, 132–153.

32. Tucson *Arizona Star,* September 9, October 11, 1931; September 11, 1934; Tucson *Citizen,* August 10, 1930; Edwin F. Carpenter, "A

Cluster of Extra-Galactic Nebulae in Cancer," ASP *Publications*, XLIII (August 1931), 247–54.

Chapter Six

1. Douglass to K.C. Babcock, December 29, 1905; January 17, 23, 1906, Box 27, Douglass Papers, UAL.
2. Babcock to Douglass, May 3, 11, 1906, Box 27; Roskruge to Douglass, May 18, 1906, Box 37, Douglass Papers, UAL.
3. Douglass to Babcock, May 8, 1906, Box 27, Douglass Papers, UAL. Douglas D. Martin, *The Lamp in the Desert: The Story of the University of Arizona* (Tucson: University of Arizona Press, 1960), 278–79.
4. Luther to Douglass, May 24, 1906, Box 7, Douglass Papers, UAL.
5. Douglass to C.W. Eliot, October 28, 1906; George W. Robinson to Douglass, November 7, 1906, Box 27, Douglass Papers, UAL.
6. Luther to Douglass, August 19, 1907; May 8, 1908, Box 7, Douglass Papers, UAL.
7. Class Book 1906/7–1907/8; Hydraulics Notes, 1907, Box 38, Douglass Papers, UAL.
8. Roskruge to Douglass, May 24, 1907, Box 37; Class Books, 1906–1913, Box 38, Douglass Papers, UAL.
9. Douglass to John M. Baer, December 26, 1906; Baer to Douglass, December 28, 1906; Campbell to Douglass, January 7, 1907; A.N. Taylor to Douglass, February 8, 1907, Box 27, Douglass Papers, UAL. Douglass to Campbell, January 11, 1907, Lick Observatory Archives.
10. Douglass to G.F. Bovard, November 2, 1907; Bovard to Douglass, November 4, 1907; George E. Hale to Douglass, February 21, 1908, Box 27, Douglass Papers, UAL.
11. Douglass to Sarah Hale Douglass, December 13, 1908, Box 148; E.S. Miller to Douglass, March 16, 1909, Box 27, Douglass Papers, UAL.
12. Travel Notes, folder 1, Box 154; Douglass to Sarah Hale Douglass, July 16, 1910, Box 148; Douglass to W.H. Pickering, September 26, 1921, Box 14, Douglass Papers, UAL.
13. Tucson *Arizona Star*, October 23, 1910.
14. Ibid., December 21, 1910.

15. Martin, *Lamp in the Desert*, 278.
16. Tucson *Arizona Star*, May 3, 17, June 1, 1911.
17. Class Book, 1910/11–1912/13; Astronomy lectures, 1908/9–1930s; Physics, 1908–1919, Box 38, "Astronomy 2," folder 5, Box 38; Douglass Papers, UAL.
18. Douglass to A.H. Wilde, December 11, 1911; to J.A.B. Scherer, December 23, 1911; Scherer to Douglass, January 9, 1912; Douglass to Willis L. Baer, December 23, 1911; to Bovard, December 23, 1911; Bovard to Douglass, January 2, 1912, Box 27, Douglass Papers, UAL.
19. Wilde to Douglass, January 12, 1912; Willis L. Moore to Douglass, January 25, 1912; John A. Miller to Douglass, February 3, 1912; E.B. Frost to Douglass, February 17, 1912, Box 27; Memo to Wilde, September 16, 1912, Box 28, Douglass Papers, UAL.
20. Diary, 1912, folder 3, Box 154; Douglass to Sarah Hale Douglass, October 30, November 17, 25, December 8, 15, 22, 1912, Box 148, Douglass Papers, UAL.
21. K.O. Bertling to Douglass, December 5, 1912; Joel Stebbins to Douglass, December 16, 1912, Box 28; Douglass to Frye, March 7, 1914, Box 7, Douglass Papers, UAL.
22. Diary and notes, 1913, folder 3, Box 154; Douglass to Sarah Hale Douglass, January 1, 1913, Box 148; Wilde to Douglass, February 11, 1913, Box 28, Douglass Papers, UAL.
23. Various material in Boxes 29–31, 34, 38–39, Douglass Papers, UAL.
24. Martin, *Lamp in the Desert*, 278–79; *1915 Desert* (Tucson: University of Arizona, 1915).
25. "Administrative Matters, January 1916," Box 28, Douglass Papers, UAL.
26. Douglass to von KleinSmid, April 19, 1916, Box 28; February 9, 1918; various letters in folder 2, Box 29, Douglass Papers, UAL.
27. Douglass to E.A. Alderman, February 25, 1911, Box 27; to Babcock, May 26, 1913; to Boardman Wright, June 15, 1914, Box 7, Douglass Papers, UAL.
28. W.E. MacGowan to Douglass, June 1, 1915; Douglass to F.H. Sykes, June 7, July 21, 22, August 14, 1915; Sykes to Douglass, July 20, 1915, Box 28; Luther to Douglass, August 7, 1915, Box 7, Douglass Papers, UAL.
29. E.C. Prentiss, "History of the Formation of the Southwestern Division of the American Association for the Advancement of

Science," ms. in folder 1; Minutes of Organization Meeting, April 10, 1920, Box 131, Douglass Papers, UAL.

30. Announcement dated October 30, 1920; Minutes, Executive Committee Meeting, December 1, 1920; Program, 1st Annual Meeting, December 2–4, 1920; Minutes, Executive Committee Meeting, December 4, 1920, folder 1, Box 131, Douglass Papers, UAL.

31. Minutes of Executive Committee Meeting, January 26, 1922; Arno B. Cammerer to Douglass, March 30, October 9, 1921, Box 131, Douglass Papers, UAL.

32. Minutes of Executive Committee Meetings, May 9, 1921; January 26, 1922; Program of 2nd Annual Meeting, January 26–28, 1922, Box 131, Douglass Papers, UAL. A.E. Douglass, "Some Aspects of the Use of the Annual Rings of Trees in Climatic Study," *The Scientific Monthly*, XV (July 1922), 5–22; *Annual Report . . . of the Smithsonian Institution . . . 1922* (Washington: Government Printing Office, 1924), 223–39.

33. Programs of Meetings, 1922–33, 1935–36, 1940, 1950, Boxes 131–32, Douglass Papers, UAL.

34. Tucson *Arizona Star*, October 5, November 5, 1921; April 10, 1933; Tucson *Citizen*, October 12, 1927. "Public Service at the Observatory, 1924–5," April 13, 1925, Box 47, Douglass Papers, UAL.

35. Tucson *Citizen*, January 9, 1924.

36. Tucson *Citizen*, January 10, 11, 1924; Tucson *Arizona Star*, January 12, 20, 1924.

37. Tucson *Citizen*, March 2, 9, 1924.

38. Ibid., May 19, 1924; Tucson *Arizona Star*, October 16, 1927.

39. Douglass to von KleinSmid, May 1, 1920; to City Council, February 8, 1922; G.H. Atchley to Douglass, March 6, 1922, Box 48; Steward Observatory Report, November 29, 1928, Box 47, Douglass Papers, UAL. Tucson *Citizen*, August 10, 1922; Tucson *Arizona Star*, February 16, 1922.

40. Lyndon R. Wilson to Douglass, January 21, 1930; Douglass to Mayor and Council of Tucson, January 27, 1931; "City Lights" notes 1931, Box 48, Douglass Papers, UAL. Tucson *Arizona Star*, January 30, 1931; Tucson *Citizen*, February 5, 1931.

41. Notes of City Council Meeting, June 21, 1932; Douglass to Mayor G.K. Smith and City Council, June 25, 1932; Edwin F. Carpenter to Douglass, July 18, 1934; Douglass to Gertrude E. Mason, July 31, 1934, Box 48, Douglass Papers, UAL.

42. "Report upon the attempted re-zoning at Campbell Avenue and Speedway, 1936 April 13;" James R. Dunseath to Douglass, April 16, 1936; Douglass to H.L. Shantz, April 20, 1936, Box 48, Douglass Papers, UAL. Tucson *Arizona Star*, April 14, 1936.
43. Douglass to von KleinSmid, May 1, 1920; Douglass to Shantz, May 10, 1930, Box 48; Steward Observatory Reports, June 4, 1924; June 3, 1925; March 27, 1927; June 12, 1929, Box 47, Douglass Papers, UAL.
44. Douglass to Shantz, January 20, February 11, 1936, Box 48, Douglass Papers, UAL.
45. Douglass to Harlow Shapley, W.S. Adams, Otto Struve, W.H. Wright (telegrams), February 28, 1936; Wright to Douglass, February 29, 1936; Shapley to Douglass, February 29, 1936; Memorandum "To the President and Governing Boards of The University," dated 1936, Box 48, Douglass Papers, UAL.
46. Department of Astronomy Report, June 12, 1929, Box 47; Steward Observatory Report, June 12, 1929, Box 47; Douglass to Shantz, May 10, 1930; Carpenter to President McCormick, December 6, 1950, Box 48, Douglass Papers, UAL.
47. Douglass to Shantz, May 9, 1930; Various notes, folders 4–5; Douglass to H.A. Spoehr, April 20, 1936; to Alfred Atkinson, October 26, 1937; Atkinson to Douglass, September 9, 1938, Box 48, Douglass Papers, UAL.

Chapter Seven

1. George Sarton, "When was tree-ring analysis discovered?" *Isis*, XLV (December 1954), 383–84; B.E. Fernow, "Age of Trees and Time of Blazing Determined by Annual Rings," United States Department of Agriculture, Division of Forestry, *Circular No. 16* (Washington, 1897), 1–6.
2. A.E. Douglass, "Some Aspects of the Use of the Annual Rings of Trees in Climatic Study," *The Scientific Monthly*, XV (July 1922), 6; A.L. Child, "Annual Growth of Trees," *Popular Science Monthly* [*PSM*], XXII (December 1882), 204–206; A.L. Child, "Concentric Rings of Trees," *PSM*, XXIV (December 1883), 259–61.
3. "Cycles: A Problem in Naming," ms. dated December 20, 1934, folder 3, Box 67; "Incidents," ms. dated 4/4/39, folder 1, Box 93, Douglass Papers, UAL. A.E. Douglass, *Climatic Cycles and Tree-Growth: A Study of the Annual Rings of Trees in Relation to*

Climate and Solar Activity (Washington: Carnegie Institution of Washington, 1919), 9.

4. *Annals of the Astrophysical Observatory of the Smithsonian Institution,* II (Washington: Government Printing Office, 1908), 1–9, 124; S.P. Langley, "On a Possible Variation of the Solar Radiation and its Probable Effect on Terrestrial Temperatures," *AJ,* XIX (June 1904), 320–21.
5. Norman and W.J.S. Lockyer, "On Solar Changes of Temperature and Variations in Rainfall in the Region Surrounding the Indian Ocean," *Proceedings of the Royal Society of London,* LXVII (1901), 410; C.G. Abbot, "The Relation of the Sun-Spot Cycle to Meteorology," *Monthly Weather Review [MWR],* XXX (April 1902), 178–81; Simon Newcomb, "A Search for Fluctuations in the Sun's Thermal Radiation through their Influence on Terrestrial Temperature," American Philosophical Society, *Transactions,* XXI (1908), 379–84.
6. A.E. Douglass, "Weather Cycles in the Growth of Big Trees," *MWR,* XXXVII (June 1909), 226; A.E. Douglass, "A Method of Estimating Rainfall by the Growth of Trees," in Ellsworth Huntington, *The Climatic Factor as Illustrated in Arid America* (Washington: Carnegie Institution of Washington, 1914), 103, 121.
7. Douglass, "Weather Cycles," 226–27.
8. Douglass, *Climatic Cycles and Tree-Growth,* 55. Notes dated 9/10/06, folder 2, Box 102, Douglass Papers, UAL.
9. A.E. Douglass, "Tree Rings and Chronology," University of Arizona *Physical Science Bulletin No. 1* (Tucson: University of Arizona, 1937), 7.
10. Douglass, "Weather Cycles," 227–28.
11. Douglass, *Climatic Cycles and Tree-Growth,* 16–17, 27–29; Douglass, "Method of Estimating Rainfall," *Climatic Factor,* 105–106. Various notes and computations, folder 3, Box 102, Douglass Papers, UAL.
12. Douglass to W.H. Pickering, March 15, 1913, Box 14, Douglass Papers, UAL. Ellsworth Huntington, "The Solar Hypothesis of Climatic Changes," Geological Society of America, *Bulletin,* XXV (November 2, 1914), 484–556, passim.
13. Douglass to H.H. Turner, April 25, 1915; to Hale, February 6, 1918; Adams to Douglass, February 27, 1918, Box 64, Douglass Papers, UAL.

14. Ellsworth Huntington, "The Secret of the Big Trees," *Harper's Monthly Magazine,* CXXV (July 1912), 292–302.
15. Raphael Zone to Douglass, September 25, October 12, 1912; A.B. Thomas to Douglass, October 30, December 10, 1912; Douglass to A. Cieslar, January 3, 1914; to H.H. Jelstrup, December 13, 1913; Freight bill from Eastern Steamship Corporation, January 21, 1913, Box 64, Douglass Papers, UAL. Douglass, *Climatic Cycles and Tree-Growth,* 23, 29–39, 114.
16. Various notes and computations, 1912–1913, Box 102; Douglass to W.H. Pickering, March 15, 1913, Box 14, Douglass Papers, UAL.
17. Douglass, *Climatic Cycles and Tree-Growth,* 45–48, 54–56. Douglass to W.B. Cannon, February 26, 1919, Box 84, Douglass Papers, UAL.
18. "Account of American Association Fund," December 16, 1918, Box 84; Douglass to Frederic E. Clements, September 14, 1918, Box 78; Douglass to E.C. Pickering, December 20, 1918, Box 14, Douglass Papers, UAL. E.C. Pickering to Douglass, April 23, 1918, Director's Files, Observatory Archives, HUA. Douglass, *Climatic Cycles and Tree-Growth,* 10–11, 49–53.
19. Douglass, *Climatic Cycles and Tree-Growth,* 49, 58, 112.
20. Ibid., 10, 66–70; Notes dated September 19, 1912; various notes dated 1912, Box 102, Douglass Papers, UAL.
21. A.E. Douglass, "Climatic Records in the Trunks of Trees," *American Forestry,* XXIII (December 1917), 733.
22. Notes and sketches dated 4/4/33, folder 1, Box 99, Douglass Papers, UAL. Arthur Schuster, "The Periodogram of Magnetic Declination as obtained from the records of the Greenwich Observatory during the years 1871–1895," Cambridge Philosophical Society, *Transactions,* XVIII (1900), 107–35; Douglass, *Climatic Cycles and Tree-Growth* (1919), 86–88.
23. A.E. Douglass, "A Photographic Periodogram of the Sun-Spot Numbers," *AJ,* XL (October 1914), 326–31.
24. A.E. Douglass, "An Optical Periodograph," *AJ,* XLI (April 1915), 173–86. Douglass to Clements, September 14, 1918, Box 78, Douglass Papers, UAL. Douglass, *Climatic Cycles and Tree-Growth* (1919), 92–96.
25. Fund Accounts, 1918, 1919, Box 81; First Typing, "Climatic Cycles and Tree-Growth," I (1919); William Barnum to Douglass, November 3, 1919, Box 108, Douglass Papers, UAL.
26. Ibid., 74–77, 80, 98–99, 101.

27. Ibid., 102–112 passim.
28. Clark Wissler to Douglass, May 22, 1914; Douglass to Wissler, June 5, 1914; Pliny E. Goddard to Douglass, June 19, 1914, Box 76, Douglass Papers, UAL.
29. Wissler to Douglass, January 3, 31, March 23, April 7, 1916; January 15, 1918, Box 76; Douglass to Ellsworth Huntington, January 25, 1916, Box 75; to Livingston Farrand, February 12, March 21, 1916, Box 64, Douglass Papers, UAL.
30. Wissler to Douglass, May 28, June 13, 1919; Douglass to Wissler, June 6, November 24, 1919, Box 76; Douglass to Earl H. Morris, August 6, 1919, Box 75; Diary, eastern trip, 1919, Box 160, Douglass Papers, UAL.
31. John C. McGregor, *Southwestern Archaeology* (Second edition, Urbana: University of Illinois Press, 1965), 33–43; Paul S. Martin and Fred Plog, *The Archaeology of Arizona: A Study of the Southwest Region* (Garden City, New York: Doubleday/Natural History Press, 1973), 10–22.
32. A.E. Douglass, "Dating Our Prehistoric Ruins," *Natural History* [*NH*], XXI (January–February 1921), 27; Douglass, *Climatic Cycles*, II, 59–60.
33. Douglass to Morris, October 12, 20, December 9, 1920, Box 75, Douglass Papers, UAL. Douglass, "Dating Our Prehistoric Ruins," 28.
34. George H. Pepper, "Pueblo Bonito," American Museum of Natural History, *Anthropological Papers*, XXVII (1920), 1; Douglass, "Dating Our Prehistoric Ruins," 28.
35. Neil M. Judd to Douglass, April 9, May 30, September 2 (telegram), 11, October 6, 1921; Douglass to Judd, May 20, August 30, September 21, 1921, Box 77, Douglass Papers, UAL. Douglass, "Dating Our Prehistoric Ruins," 27; Paul H. Oehser, ed., *National Geographic Society Research Reports, 1890–1954* (Washington: National Geographic Society, 1975), 187.
36. Douglass to Clements, June 4, 1922, Box 80, Douglass Papers, UAL.
37. Neil M. Judd, "The Pueblo Bonito Expedition of the National Geographic Society," *The National Geographic Magazine [NGM]*, XLI (March 1922), 323.
38. Judd to Douglass, June 3, 27, 1922; Douglass to Judd, June 14, August 1, 1922, Box 77, Douglass Papers, UAL.
39. Douglass to Judd, August 1, September 22, 1922; Judd to Douglass, August 5, 1922, Box 77, Douglass Papers, UAL.

Douglass, "The Secret of the Southwest Solved by Talkative Tree Rings," *NGM*, LVI (December 1929), 764.

40. Douglass, "Evidence of Climatic Effects in the Annual Rings of Trees," *Ecology*, I (January 1920), 24–32.
41. Ibid., 34–35. Douglass to G.A. Pearson, September 24, 1920, Box 76, Douglass Papers, UAL.
42. G.E. Armstrong to Douglass, January 17, 1921; Douglass to Armstrong, January 25, 1921; Douglass to Walter Nordhoff, February 4, 1921; to Robert G. Sproul, February 10, 1921; to R.E. Burton, March 7, December 14, 1921, Box 126, Douglass Papers, UAL. Douglass, *Climatic Cycles*, II, 55, 91.
43. Douglass, *Climatic Cycles*, I (1919), 101–102; II (1928), 125–126; Douglass, "Some Topographic and Climatic Characters in the Annual Rings of the Yellow Pines and Sequoias of the Southwest," APS *Proceedings*, LXI (1922), 120–121. E. Walter Maunder to Douglass, February 18, 1922; Douglass to Maunder, March 23, 1922, Box 64, Douglass Papers, UAL. For a good overview of Maunder's work, see John A. Eddy, "The Maunder Minimum," *Science*, June 18, 1976, 1189–1202.
44. Douglass to Judd, June 4, 1924, Box 77, Douglass Papers, UAL. Douglass, *Climatic Cycles*, II, 86.
45. Douglass, *Climatic Cycles*, II, 19, 52–54, 61–63, 87–90. Annual Report to Carnegie Institution, September 4, 1924, folder 4, Box 79, Douglass Papers, UAL.
46. "Big Auto Trip," 1925, diary, folder 2, Box 160, Douglass Papers, UAL. Douglass, *Climatic Cycles*, II, 55–56, 63–64, 77, 85, 91, 92.
47. Douglass to P.W. Weirick, August 24, 1925, Box 65, Douglass Papers, UAL. Douglass, *Climatic Cycles*, II, 24–26, 54, 88–89.
48. "Present Status of Tree Ring Studies," ms. dated May 7, 1927; J. Arthur Harris to Douglass, December 30, 1925, Box 65, Douglass Papers, UAL. J. Arthur Harris, "The Correlation Between Sun-Spot Number and Tree Growth," *MWR*, LIV (January 1926), 13–14.
49. Douglass, "Solar Records in Tree Growth," *Science*, March 4, 1927, 220–221.
50. Douglass to Clements, August 28, 1926, Box 80, Douglass Papers, UAL. Douglass, *Climatic Cycles*, II, 38–40, 43.
51. John C. Merriam to Douglass, November 6, 1928, Box 78, Douglass Papers, UAL. "The Second Conference on Cycles," *Geographical Review*, XIX (April 1929), 296–306.

52. Douglass to Merriam, May 7, 1927, Box 78; to Clements, December 10, 1926; March 24, 1927; Clements to Douglass, September 14, 1924; November 25, 1925; April 24, June 30, 1926; April 26, May 5, 1927, Box 80; Annual Report to the Carnegie Institution, September 3, 1926, folder 4, Box 79; Notes dated November 30, 1923; William S. Gilbert to Douglass, June 11, 1927, Box 109, Douglass Papers, UAL.
53. Douglass, *Climatic Cycles*, II, 8–13, 31, 34, 56–57, 63, 92–93, 96–97.
54. Ibid., 67–68.
55. Ibid., 100–138 passim.
56. Douglass to Merriam, September 17, 1929, Box 78; to Morris, October 29, 1929, Box 75; to Judd, November 27, 1929, Box 77; "Conference With J.C.M.," notes dated December 1929–January 1930, folder 4, Box 78, Douglass Papers, UAL.
57. Lecture Report, July 1, 1929 to June 30, 1930, folder 2, Box 136, Douglass Papers, UAL. Douglass, "Secret of the Southwest," 740.

Chapter Eight

1. Douglass to A.V. Kidder, August 1, 22, 1922; January 15, 1923, Box 75; Douglass to Clements, October 28, 1922, Box 80; Gilbert H. Grosvenor to Douglass, December 15, 1922, Box 77, Douglass Papers, UAL.
2. Judd to Douglass, April 24, 1923, Box 77; Note dated October 25, 1923, folder 1, Box 73, Douglass Papers, UAL. Douglass, "Secret of the Southwest," *NGM*, LVI, 745, 750; Douglass, *Climatic Cycles*, II, 61.
3. Judd to Douglass, September 6, November 24, 1924; Grosvenor to Douglass, November 15, 1924, Box 77, Douglass Papers, UAL.
4. Grosvenor to Douglass, June 25, 1926; Douglass to Judd, August 28, 1926, Box 77; National Geographic Society Fund Accounts, 1923–28, folder 2, Box 84, Douglass Papers, UAL. Douglass, *Climatic Cycles*, II, 60, 65–66.
5. Judd to Douglass, September 23, 1926; May 3, 1927; Douglass to Judd, March 24, April 29, May 6, June 24, 1927, Box 77, Douglass Papers, UAL. Douglass, "Dating Pueblo Bonito and Other Ruins of the Southwest," National Geographic Society, *Contributed Technical Papers [CTP]*, Pueblo Bonito Series, Number 1 (Washington: National Geographic Society, 1935), 21.

6. Douglass, "Dating Pueblo Bonito," 21, 52; Douglass, "Secret of the Southwest," 750. Douglass to Judd, July 14, 1927, Box 77, Douglass Papers, UAL.
7. Douglass to Judd, August 20, 1927, Box 77, Douglass Papers, UAL. Douglass, "Dating Pueblo Bonito," 21–22; Douglass, "Secret of the Southwest," 750.
8. Kidder to Douglass, July 3, 1927, Box 75; "Archaeological Side of Tree Rings and Climatic Records," ms. in folder 9, Box 135, Douglass Papers, UAL. A.V. Kidder, "Southwestern Archaeological Conference," *Science*, November 18, 1927, 489–91; McGregor, *Southwestern Archaeology*, 62–73.
9. Morris to Douglass, November 6, 1927, Box 75; Douglass to Judd, December 3, 1927, Box 77, Douglass Papers, UAL. Douglass, "Dating Pueblo Bonito," 22.
10. Douglass to Morris, January 19, 1928, Box 77, Douglass Papers, UAL.
11. Douglass to Judd, February 3, 13, 1928; Judd to Douglass, February 25, March 2, 1928, Box 77, Douglass Papers, UAL. Douglass, "Dating Pueblo Bonito," 22.
12. Douglass to Grosvenor, February 27, 1928; Douglass to Judd, August 21, 1928, Box 77, Douglass Papers, UAL. Interview with Lyndon L. Hargrave, July 16, 1976, Prescott, Arizona.
13. Douglass to Judd, April 18, 22, August 21, 1928, Box 77, Douglass Papers, UAL.
14. Douglass to Judd, April 22, 1928, Box 77, Douglass Papers, UAL. Douglass, "Secret of the Southwest," 751–52.
15. Douglass, "Secret of the Southwest," 753, 756–68.
16. Douglass to Judd, June 12, 1928, Box 77, Douglass Papers, UAL.
17. Douglass to Judd, August 20, 21, 31, 1928, Box 77, Douglass Papers, UAL. Douglass, "Secret of the Southwest," 757, 762–63.
18. Judd to Douglass, September 5, 1928, Box 77, Douglass Papers, UAL. Douglass, "Dating Pueblo Bonito," 26; Douglass, "Secret of the Southwest," 762–63.
19. Interview with Emil W. Haury, March 31, 1977, Tucson, Arizona. Diary, "Eastern Trip," 1928–29, folder 3, Box 160; Douglass to Grosvenor, March 2, 1929; to Judd, March 31, April 9 (telegram), 1929, Box 77; Harold S. Colton to Douglass, April 17, 1929 (telegram), Box 73, Douglass Papers, UAL.
20. Interview with Emil W. Haury, March 31, 1977. Diary, 1929, folder 3, Box 160, Douglass Papers, UAL. Emil W. Haury and

Lyndon L. Hargrave, "Recently Dated Pueblo Ruins in Arizona," *Smithsonian Miscellaneous Collections [SMC]*, LXXXII, No. 11 (August 18, 1931), 2–5.

21. Douglass, "Secret of the Southwest," 764–65; Haury and Hargrave, "Recently Dated Pueblo Ruins," 9–10, 25, 30.
22. Emil W. Haury to Judd, June 13, 1929, Box 77, Douglass Papers, UAL. Emil W. Haury, "HH-39: Recollections of a Dramatic Moment in Southwestern Archaeology," *Tree-Ring Bulletin [TRB]*, XXIV (May 1962), 12–13; Haury and Hargrave, "Recently Dated Pueblo Ruins," 13–14.
23. Douglass, "Secret of the Southwest," 766–70; Douglass, "Dating Pueblo Bonito," 41.
24. Douglass to Judd, November 19, 1929, Box 77; Haury to Douglass June 29, July 10, 1929; Douglass to Hargrave, November 19, 1929, Box 75, Douglass Papers, UAL. Haury and Hargrave, "Recently Dated Pueblo Ruins," 6, 80–95, 116.
25. National Geographic Society Fund Accounts, December 1928–June 1930, folder 3, Box 84, Douglass Papers, UAL.
26. Judd to Douglass, September 19, October 11, 19, 24, 1929, Box 77; Douglass to A.L. Kroeber, September 17, 1931, Box 75, Douglass Papers, UAL.
27. Judd to Douglass, October 24, 1929, Box 77; Paul S. Martin to Douglass, December 5, 1929, Box 65; Various letters in folder 5, Box 65, Douglass Papers, UAL.
28. Interview with Emil W. Haury, March 31, 1977. Douglass, "Tree Rings and their Relation to Solar Variations and Chronology," *Annual Report...of the Smithsonian Institution...1931* (Washington: Government Printing Office, 1932), 311.
29. Douglass to Colton, March 20, 1933, Box 73, Douglass Papers, UAL. Douglass to Colton, May 13, 20, 1931; May 8, 13, 20, 1933; to John C. McGregor, May 8, 1933, Museum Archives, Museum of Northern Arizona [MNA], Flagstaff.
30. W.S. Stallings, Jr., "A Tree-Ring Chronology for the Rio Grande Drainage in Northern New Mexico," National Academy of Sciences, *Proceedings*, XIX (1933), 805. Douglass to Fred W. Emerson, June 3, 1931, Box 66; W.S. Stallings, Jr., to Director, Laboratory of Anthropology, Inc., Santa Fe, October 15, 1934, Box 76, Douglass Papers, UAL.
31. Florence M. Hawley, "The Significance of the Dated Prehistory of Chetro Ketl, Chaco Canyon, New Mexico," University of New

Mexico *Bulletin 246* (Albuquerque: University of New Mexico, 1934), 5. Douglass to Perry R. Taylor, August 3, 1934, Box 67; Florence M. Hawley to Douglass, March 21, September 30, 1935, Box 75, Douglass Papers, UAL.

32. Various drafts, Box 124; Judd to Douglass, April 21, July 25, 1933, Box 77, Douglass Papers, UAL.
33. Various drafts, Box 124; Judd to Douglass, December 7, 1933; July 6, 1935; Douglass to Judd, February 24, August 25, 1934; June 4, 1935, Box 77; Frederick G. Vosburgh to Douglass, February 1, 1935, Box 124, Douglass Papers, UAL.
34. Douglass, "The Central Pueblo Chronology," *TRB*, II (April 1936), 29–31. Douglass to Jesse L. Nusbaum, April 29, 1930, Box 75, Douglass Papers, UAL.
35. Douglass, "Central Pueblo Chronology," 31; Douglass, "Climatological Researches," *CIW Year Book No. 31* (1931–32), 218.
36. Douglass to Morris, November 10, 21, 1931, Box 75, Douglass Papers, UAL.
37. Douglass, "Typical Ring-Record from Chaco Canyon, 700 to 850, CK-331," *TRB*, III (January 1937), 20; Douglass, "Central Pueblo Chronology," *TRB*, II, 31; Douglass, "Climatological Researches," *CIW Year Book No. 32* (1932–33), 211.
38. Douglass to Morris, January 10, March 17, 1933; Morris to Douglass, January 16, March 20, 1933, Box 75, Douglass Papers, UAL. Douglass, "Tree Rings and Chronology," UA *Physical Science Bulletin No. 1* (1937), 12.
39. Douglass to Morris, April 10, June 6, 11, 1933, Box 75; to Judd, May 1933, Box 77, Douglass Papers, UAL. Douglass, "Climatic Researches," *CIW Year Book No. 33* (1933–34), 198.
40. Douglass, "Central Pueblo Chronology," *TRB*, II, 29–32; Douglass, "Climatological Researches," *CIW Year Book No. 34* (1934–35), 219; *No. 35* (1935–36), 228. Douglass to Alfred Peterson, August 24, 1935, Box 67, Douglass Papers, UAL.

Chapter Nine

1. List of scientific meetings attended since 1930, dated 12/12/35, Box 34; Douglass to Merriam, October 10, December 21, 1930, Box 78, Douglass Papers, UAL. Douglass, "Climatological Researches," *CIW Year Book No. 31* (1931–32), 217; *No. 32* (1932–33), 208–11.

2. Douglass, "Evidences of Cycles in Tree Ring Records," National Academy of Sciences, *Proceedings*, XIX (1933), 350–60; "Symposium on Climatic Cycles," ibid., 349–88.
3. Various notes in folder 9, Box 78; "Tree Ring Cycles and Their Solution," ms. dated April 30–May 22, 1932, folder 5, Box 66; Douglass to Spoehr, June 15, 1933, Box 80, Douglass Papers, UAL.
4. Notes dated August 31, 1942, folder 5, Box 69, Douglass Papers, UAL.
5. Douglass to William J. Showalter, October 24, 1930, Box 33; Showalter to Douglass, December 12, 1930, Box 66, Douglass Papers, UAL.
6. Various material, Box 94; Douglass to Merriam, March 20, April 10, 1933, Box 78; to Nusbaum, March 20, 1933, Box 75, Douglass Papers, UAL.
7. Various material, Box 94; "Tree Ring Research," ms. dated 9/25/33, folder 1, Box 67; Douglass to Merriam, September 25, 1933, Box 78, Douglass Papers, UAL.
8. "The Carnegie Cooperation," ms. dated 5/25/36, folder 9; Merriam to Douglass, January 5, 1935; Douglass to Merriam, January 19, 23, 1935, Box 78, Douglass Papers, UAL.
9. Merriam to Douglass, January 26, February 1, 1935, Box 78, Douglass Papers, UAL.
10. Seely to Douglass, March 2, 1935, Box 67; Colton to Shantz, February 14, 1935, Box 73, Douglass Papers, UAL. Tucson *Arizona Star*, February 9, 1935; "Tree Rings & Weather," *Time* February 18, 1935, 46–47.
11. Various notes, folder 1, Box 111, Douglass Papers, UAL.
12. Douglass to Spoehr, November 26, 1935, Box 80; to Frank F. Bunker, March 10, 14, April 3, 1936; to S.B. Nicholson, December 10, 1935; to Dinsmore Alter, December 10, 1935; to Charles G. Abbot, March 14, 1936, Box 115; to Merriam, December 21, 1935, January 3, 1936; Merriam to Douglass, February 27, December 8, 1936, Box 78; Various drafts, Boxes 112–114, Douglass Papers, UAL.
13. Douglass, *Climatic Cycles and Tree-Growth, Volume III: A Study of Cycles* (Washington: Carnegie Institution of Washington, 1936), 141.
14. Douglass to E.L. Neill, August 4, 1932; Notes dated 9/25/33, Box 66; Douglass to Clarence G. White, July 25, 1933, Box 76;

Douglass to von KleinSmid, September 25, 1933, Box 37, Douglass Papers, UAL.

15. Various notes and sketches; Douglass to Paul S. Burgess, February 18, 1937, Box 86; to Merriam, February 24, 1937, Box 78, Douglass Papers, UAL.
16. Douglass to Atkinson, October 25, 1937, Box 93; Atkinson to Douglass, December 6, 1937; Various notes and memoranda, Box 86, Douglass Papers, UAL. "Tree-Ring Laboratory," *TRB*, IV (January 1938), 2.
17. Various notes in folder 3, Box 86; Budget, 1938–39, folder 2, Box 88; Douglass to R. MacClellan Brady, January 8, 1938, Box 8; Douglass to Judd, February 25, 1938, Box 77; Various notes, Box 93, Douglass Papers, UAL.
18. "Inauguration of Alfred Atkinson, M.S., D.Sc. as President of the University," *General Bulletin No. 4* (Tucson: University of Arizona, 1939), 57–58; Phoenix *Arizona Republic*, April 13, 1938.
19. Douglass to Warren Weaver, April 19, 1938, Box 93; Merriam to Douglass, June 7, 1938; Douglass to Merriam, July 6, 1938, Box 78; Notes dated 7–10–38; Application for WPA assistance, dated 11–15–38, folder 4, Box 86, Douglass Papers, UAL.
20. Glock, "Report on the First Tree Ring Conference," *TRB*, I (July 1934), 4–5; Douglass, "Editorial," ibid., 3. Douglass to Stallings, August 1, 1934, Box 76, Douglass Papers, UAL.
21. Getty, "Second Annual Tree Ring Conference," *TRB*, II (July 1935), 4–5; "Third Annual Tree Ring Conference," ibid., III (July 1936), 3–4; "Fourth Annual Tree Ring Conference," ibid., IV (July 1937), 2; Emil W. Haury, "Editorial," ibid., IV (October 1937), 2.
22. C.W. Ceram [Kurt W. Marek], *The First American: A Study of North American Archaeology* (New York: Harcourt Brace Jovanovich, Inc., 1971), 167; Douglass, "Accuracy in Dating—II," *TRB*, I (January 1935), 20. Interview with Emil W. Haury, March 31, 1977. Douglass to H.S. Gladwin, December 9, 1934, Box 74, Douglass Papers, UAL.
23. Draft of proposed letter to Gladwin, dated 1–6–40, folder 4, Box 74, Douglass Papers, UAL.
24. Harold S. Gladwin, "Methods and Instruments for Use in Measuring Tree-Rings," *Medallion Papers No. XXVII* (Globe: Gila Pueblo, 1940), 1.
25. Ibid., 6–10.

26. Harold S. Gladwin, "Tree-Ring Analysis: Methods of Correlation," *Medallion Papers No. XXVIII* (Globe: Gila Pueblo, 1940), 4.
27. Ibid., 11–61.
28. Douglass to Gladwin, May 9, 1940, Box 74, Douglass Papers, UAL. Edmund Schulman, "Variations Between Ring Chronologies in and near the Colorado River Drainage Area," *TRB*, VIII (April 1942), 26-32.
29. Harold S. Gladwin, "A Review and Analysis of the Flagstaff Culture," *Medallion Papers No. XXXI* (Globe: Gila Pueblo, 1943), 55.
30. Ibid., 56–58, 65–68.
31. McGregor to Douglass, November 13, 1943, Box 75; E.S. Schulman to Douglass, February 15, March 8, 1944, Box 90, Douglass Papers, UAL.
32. Harold S. Gladwin, "Tree-Ring Analysis. Problems of Dating—I: The Medicine Valley Sites," *Medallion Papers No. XXXII* (Globe: Gila Pueblo, 1944), 2, 43.
33. Douglass to McGregor, November 30, 1945, Box 75; Colton to Douglass, December 13, 1945, Box 73, Douglass Papers, UAL. Harold Sellers Colton, *The Sinagua: A Summary of the Archaeology of the Region of Flagstaff, Arizona* (Flagstaff: Northern Arizona Society of Science and Art, 1946), 33, 305.
34. Colton, *The Sinagua*, 33–34.
35. Ibid., 34.
36. Douglass to Abbot, October 11, 1946, Box 70, Douglass Papers, UAL. Douglass to Colton, December 20, 1946, January 22, 1947, Manuscript Collection, MNA.
37. Various notes and drafts, folders 4–5, Box 123, Douglass Papers, UAL. Douglass, "Precision of Ring Dating in Tree-Ring Chronologies," *University of Arizona Bulletin*, XVII (Tucson: University of Arizona, 1947), 9–10, 16–20.
38. Interview with Emil W. Haury, March 31, 1977. H.S. Gladwin, "Tree-Ring Analysis: Tree-Rings and Droughts," *Medallion Papers No. XXXVII* (Globe: Gila Pueblo, [1947]), 2–16, 19–33.
39. Gladwin, "Tree-Rings and Droughts," 9.

Chapter Ten

1. "Report on Observatory Sites," January 1, 1944; Douglass to Carpenter, January 11, June 10, 1944; to Atkinson, February 14,

1944; "Observatory Site Report," June 7, 1944, Box 48, Douglass Papers, UAL.

2. Various material, Box 48, Douglass Papers, UAL.
3. "Dramatic Incidents in Tree-Ring Studies," ms. dated 4–18–49, folder 1, Box 93; George R. McEwen to Lassetter, July 14, 1941; Douglass to McEwen, August 5, 1941; to Schulman, September 9, 1941; Agreement dated September 23, 1941, in folder 2, Box 92; "Conferences," 1941-42, folder 5, Box 69, Douglass Papers, UAL.
4. Douglass to Roy Martindale, October 8, November 12, 1941; Schulman to Douglass, October 8 (telegram), 13, 20, 25 (telegram), 1941; Douglass to Atkinson, October 31, 1941, Box 92, Douglass Papers, UAL.
5. Various reports by Schulman; Douglass to Martindale, May 14, 1942; Schulman to Douglass, May 22, 1942 (telegram), Box 92; "Dramatic Incidents in Tree-Ring Studies," ms. dated 4–18–49, folder 1, Box 93, Douglass Papers, UAL. Sibin Pinkayan, "Areal Distribution of Wet and Dry Years," Ph.D. dissertation, Colorado State University, 1965, 49, 74–77.
6. Report, Laboratory of Tree-Ring Research, 1941–42, folder 8, Box 87; Schulman to Martindale, June 12, 1942; Martindale to Douglass, August 6, 1942, Box 92, Douglass Papers, UAL.
7. Douglass to F.W. Reichelderfer, August 12, 1943, Box 70; Report, Laboratory of Tree-Ring Research, 1943–44, Box 87; Atkinson to American Philosophical Society, March 9, 1945, Box 84, Douglass Papers, UAL.
8. Douglass to Martindale, July 24, 1942, Box 92; to Grosvenor, October 31, 1942, Box 77; to Merriam, August 24, 1943, Box 78; to H.J. Bauer, May 4, 1944; to Henry A. Wallace, November 1, 1943; Charles F. Sarle to Douglass, August 28, 1943, Box 70; L.K. Doutrick to Douglass, June 8, 1944, Box 84, Douglass Papers, UAL.
9. University of Arizona Gift Report, January 5, 1949, folder 12, Box 84; Report, Laboratory of Tree-Ring Research, 1949–50, folder 10, Box 87; Report ms. dated 3–31–50 (by Schulman), folder 8, Box 92; Douglass to Schulman, February 3, 1950, Box 132, Douglass Papers, UAL.
10. Various reports and budget material, Boxes 87–89; Memorandum dated 4–2–47, folder 8, Box 86; Plans dated January 1949, folder 1, Box 71, Douglass Papers, UAL.
11. Notes dated June 9, 1957, folder 8, Box 155; Various material in folder 6, Box 130; Douglass to Moses H. Douglass, January 29, 1947, Box 149, Douglass Papers, UAL.

12. Douglass to Fred L. Whipple, May 10, 1954, Box 71, Douglass Papers, UAL.
13. Ernest W. Brown, "A Possible Explanation of the Sunspot Period," *Monthly Notices of the Royal Astronomical Society*, LX (Supplementary Number, 1900), 599–606.
14. Inigo Jones to Douglass, April 3, 1930; Inigo Jones, *My "Nephelo-Coccygia,"* Box 75; Carl Farseth to Douglass, March 25, 1936; Larry F. Page to Douglass, March 8, 1937, Box 68, Douglass Papers, UAL.
15. Douglass to Whipple, May 10, 1954, Box 71; to Atkinson, May 24, 1946; to J. Byron McCormick, February 3, 1948; "Sunspot Periodicities," dated 3–3–50, Box 86; "Survey of Cycle Studies," ms. dated 11–29–51, folder 1, Box 93, Douglass Papers, UAL.
16. Douglass to Atkinson, May 24, 1946, Box 86; Notes dated May 18, 1944, folder 3, Box 70, Douglass Papers, UAL.
17. Douglass to Atkinson, May 24, 1946; "Sunspot Periodicities," dated 3–3–50, Box 86, Douglass Papers, UAL.
18. Douglass to Atkinson, May 24, December 5, 1946; "Planetary Hypothesis of Climatic Changes," ms. dated 11–7–46, Box 86; Douglass to Whipple, May 10, 1954, Box 71, Douglass Papers, UAL.
19. "Planetary Hypothesis," ms. dated 11–7–46, Box 86, Douglass Papers, UAL.
20. "Short Cyclics and Inner Planets," ms. dated 7–4–52; Notes dated June 11, 1950, Box 106; "Sunspot Periodicities," dated 3–3–50, Box 86; "Memorandum Regarding Status of Climatic Study," ms. dated September 16, 1951, Box 93, Douglass Papers, UAL.
21. Douglass to John W. Mauchley, March 31, 1952, Box 71; "Survey of Cycle Work," ms. dated 10–9–55, Box 87, Douglass Papers, UAL. Douglass, "Climatic Cycles in the Southwestern United States: Based on Observations of the Sun, Planets and Tree Rings," ms. in files of Bryant Bannister, Laboratory of Tree-Ring Research, University of Arizona, Tucson.
22. Douglass to Vannevar Bush, May 21, 24, June 1, 1954, Box 117; to Whipple, May 10, 1954, Box 71, Douglass Papers, UAL.
23. Douglass to Haury, December 28, 1956, Box 75; Sketch dated 10–8–56, folder 4, Box 107, Douglass Papers, UAL. Douglass to Colton, June 7, 1956, Manuscript Collection, MNA.
24. Douglass to Bush, November 15, 1954; January 11, 1956, Box 79; to President, Carnegie Institution of Washington, May 21, 1956, Box 87; Drafts and discards, folder 2, Box 117, Douglass Papers, UAL.

25. Colton to Douglass, January 12, March 21, 1950; memorandum dated January 1950; Douglass to Colton, January 25, 1950 (telegram), Manuscript Collection, MNA.
26. Douglass to Stallings, January 17, 1950, Box 76; to McCormick, August 1, 1950, Box 86; Report of Terah L. Smiley, 1950–51; Report, Lab of TR Research, June 9, 1951, folder 11, Box 87; Harrison Brown to Douglass, December 16, 1957, Box 71, Douglass Papers UAL.
27. Various budget material, Box 89, Douglass Papers, UAL.
28. Schulman to Douglass, January 12, 1953; Report, Lab of TR Research, 1955–56, Box 87, Douglass Papers, UAL. Edmund Schulman, "Bristlecone Pine, Oldest Known Living Thing," *NGM*, CXIII (March 1958), 358–60.
29. Alan T. Waterman to Richard A. Harvill, January 18, 1956; Report, Lab of TR Research, 1956–57, Box 87, Douglass Papers, UAL. Schulman, "Bristlecone Pine," *NGM* 355–66.
30. Douglass to Harvill, March 3, 1958; February 24, 1959; Herbert D. Rhodes to Harvill, February 28, 1958, Box 87, Douglass Papers, UAL. "Edmund Schulman, 1908–1958," *TRB*, XXII, 2.
31. Douglass, "Climatic Cycles in the Southwestern United States," Bannister files, Tree-Ring Lab. Notes dated May 27–8, 1957, folder 6, Box 87, Douglass Papers, UAL.
32. Various material, folder 6, Box 140; Notes dated June 9, 1957, folder 8; Douglass to Harvill, January 7, 1957, Box 155; October 29, 1959, Box 87, Douglass Papers, UAL. Ida W. Douglass to Colton, December 20, 1959; May 20, 1961, Manuscript Collection, MNA.
33. Interview with Bryant Bannister, February 21, 1978, Tucson, Arizona. Ida W. Douglass to Colton, August 28, 1961, Manuscript Collection, MNA.
34. Colton to Ida W. Douglass, March 22, 1962, Manuscript Collection, MNA. Tucson *Arizona Star*, March 21, 1962; *New York Times*, March 21, 1962; London *Times*, March 25, 26, 1962. "Death and Obituary Notices," Box 155, Douglass Papers, UAL. Charles W. Ferguson, "The Tree-Ring Society: May 1962 Meetings," *TRB*, XXV (December 1962), 11.

Selected Bibliography

MANUSCRIPT COLLECTIONS

California Institute of Technology Archives (Pasadena, California)
  George Ellery Hale Papers
Harvard University Archives (Cambridge, Massachusetts)
  Observatory Archives
    Director's Files
    E.C. Pickering Papers
Lowell Observatory Archives (Flagstaff, Arizona)
  A.E. Douglass Papers
  Letter Books
  Percival Lowell Papers
  W.L. Putnam Correspondence
  Miscellaneous Papers
Museum of Northern Arizona (Flagstaff, Arizona)
  Colton Library
    Manuscript Collection
    Museum Archives
The University of Arizona (Tucson, Arizona)
  Special Collections, University Library
  Andrew Ellicott Douglass Papers
University of California (Santa Cruz, California)
  Mary Lea Shane Archives of Lick Observatory

PERIODICALS

Boston Evening Transcript, 1900.
Flagstaff *Coconino Sun*, 1898–1906.
Flagstaff *Coconino Weekly Sun*, 1894–1896.
Flagstaff *Sun-Democrat*, 1897.
New York Herald, 1901.
New York Times, 1901–1962.
Phoenix *Arizona Gazette*, 1894.
Phoenix *Arizona Republic*, 1930–1962.
Phoenix *Arizona Republican*, 1894–1930.
Prescott *Arizona Journal-Miner*, 1894.
Tucson *Arizona Star*, 1894–1962.
Tucson *Arizona Weekly Citizen*, 1894.
Tucson *Citizen*, 1894–1962.

Books and Articles

Abbot, Charles G. "The Relation of the Sun-Spot Cycle to Meteorology," *Monthly Weather Review*, XXX (April, 1902), 78–81.

______. "The Solar Constant of Radiation," *Science*, March 6, 1914, 335–48.

Abell, George. *Exploration of the Universe*. Brief edition. New York: Holt, Rinehart and Winston, 1969.

Annals of the Lowell Observatory. I (Boston: Houghton, Mifflin and Company, 1898); II (Cambridge: The University Press, 1900); III (privately published, 1905).

Bailey, Solon I. *The History and Work of Harvard Observatory, 1839 to 1927*. New York: McGraw-Hill Book Company, Inc., 1931.

Ball, Robert S. "Mars," Astronomical Society of the Pacific, *Publications*, V (January 28, 1893), 23–36.

Bannister, Bryant. "Andrew Ellicott Douglass (1867-1962)," *Year Book of the American Philosophical Society* (1965), 121–25.

Barnard, E.E. "On the Third and Fourth Satellites of Jupiter," *Astronomische Nachrichten*, CXLIV (1897), 321–30.

Bates, Henry H. "The Chemical Constitution of *Mars'* Atmosphere," Astronomical Society of the Pacific, *Publications*, VI (December 1, 1894), 300–302.

Berendzen, Richard, and others. *Man Discovers the Galaxies*. New York: Science History Publications, 1976.

Berry, Arthur. *A Short History of Astronomy: From Earliest Times Through the Nineteenth Century.* New York: Dover Publications, Inc., 1961; orig. pub. 1898.

Brown, Ernest W. "A Possible Explanation of the Sun-spot Period," *Monthly Notices of the Royal Astronomical Society,* LX (Supplementary Number, 1900), 599–606.

Burns, George P. "Rainfall and Width of Annual Rings in Vermont Forests," Vermont Agricultural Experiment Station, *Bulletin 298* (June 1929).

Campbell, W.W. "An Explanation of the Bright Projections Observed on the Terminator of *Mars,*" Astronomical Society of the Pacific, *Publications,* VI (March 31, 1894), 103–12.

______."The Spectrum of Mars," Astronomical Society of the Pacific, *Publications,* VI (August 15, 1894), 228–36.

______."Concerning an Atmosphere on *Mars,*" Astronomical Society of the Pacific, *Publications,* VI (December 1, 1894), 273–83.

______."*Mars,* By Percival Lowell," Astronomical Society of the Pacific, *Publications,* VIII (August 1, 1896), 207–20.

Carpenter, Edwin. F. "A Cluster of Extra-Galactic Nebulae in Cancer," Astronomical Society of the Pacific, *Publications,* XLIII (August, 1931), 247–54.

Ceram, C.W. (Kurt W. Marek) *The First American: A Story of North American Archaeology.* New York: Harcourt Brace Jovanovich, Inc., 1971.

Child, A.L. "Annual Growth of Trees," *Popular Science Monthly,* XXII (December, 1882), 204–206.

______."Concentric Rings of Trees," *Popular Science Monthly,* XXIV (December, 1883), 259–61.

Chriss, Michael. "The Stars Move West: The Founding of the Lick Observatory," *Mercury,* II (July/August, 1973), 10–15.

Cogshall, W.A. "A Trip to the Summit of Orizaba," *Appalachia,* VIII (March, 1898), 350–55.

Colton, Harold S. *The Sinagua: A Summary of the Archaeology of the Region of Flagstaff, Arizona.* Flagstaff: Northern Arizona Society of Science and Art, 1946.

Davis, Francis, "Rainfall a Factor of Tree Increment," *Forestry Quarterly,* X (June, 1912), 222–28.

Douglass, A.E. "Accuracy in Dating—II," *Tree-Ring Bulletin,* I (January, 1935), 19–21.

______."The Altitudes of Orizaba and Popocatepetl," *Appalachia,* VIII (March 1898), 356–61.

______."An Ascent of Popocatepetl," *Popular Astronomy*, V (February, 1898), 505–508.

______."The Astronomer's Globe," *Popular Astronomy*, V (June, 1897), 57–62.

______."Atmosphere, Telescope and Observer," *Popular Astronomy*, V (June, 1897), 64–84.

______."The Callendar Sunshine Recorder and Some of the World-Wide Problems to Which This Instrument Can Be Applied," *Proceedings of the Second Pan American Scientific Congress*, II (1917), 570–79.

______."The Central Pueblo Chronology," *Tree-Ring Bulletin*, II (April, 1936), 29–34.

______.*Climatic Cycles and Tree-Growth*. 3 vols. Washington: Carnegie Institution of Washington, 1919–1936.

______."Climatic Records in the Trunks of Trees," *American Forestry*, XXIII (December, 1917), 732–35.

______."Climatological Researches," Carnegie Institution *Year Book No. 31* (1931–32), 217–19.

______."Climatological Researches," Carnegie Institution *Year Book No. 32* (1932–33), 208–11.

______."Climatological Researches," Carnegie Institution *Year Book No. 33* (1933–34), 196–98.

______."Climatological Researches," Carnegie Institution *Year Book No. 34* (1934–35), 215–19.

______."Climatological Researches," Carnegie Institution *Year Book No. 35* (1935–36), 227–30.

______."Climatological Researches," Carnegie Institution *Year Book No. 36* (1936–37), 228–30.

______."A Cloud-Like Spot on the Terminator of Mars," *Astrophysical Journal*, I (February, 1895), 127–30.

______."A Combination Telescope and Dome," *Astrophysical Journal*, I (May, 1895), 401–10.

______."Dating Our Prehistoric Ruins," *Natural History*, XXI (January–February, 1921), 27–30.

______."Dating Pueblo Bonito and Other Ruins of the Southwest," National Geographic Society *Contributed Technical Papers*, Pueblo Bonito Series, No. 1, 1935.

______."Drawings of Comet *a*, 1910," *Popular Astronomy*, XVIII (March, 1910), 162–63.

______."Drawings of Jupiter's Third Satellite," *Astronomische Nachrichten*, CXLIII (1897), 363–66.

______."Editorial," *Tree-Ring Bulletin*, I (July, 1934), 2–3.

______."The Effect of Mountains on the Quality of the Atmosphere," *Popular Astronomy*, VII (August, 1899), 354–65.

______."Effects of High-Mountain Climbing," *Appalachia*, VIII (March, 1898), 361–67.

______."Estimated Tree-Ring Chronology: 150–300 A.D." *Tree-Ring Bulletin*, V (January, 1939), 18–20.

______."Evidence of Climatic Effects in the Annual Rings of Trees," *Ecology*, I (January, 1920), 24–32.

______."Evidences of Cycles in Tree Ring Records," National Academy of Sciences, *Proceedings*, XIX (March, 1933), 350–60.

______."The First Satellite of Jupiter," *Astronomische Nachrichten*, CXLVI (1898), 346–55.

______."Gegenschein Observations," *Popular Astronomy*, II (September, 1894), 29–32.

______."General Methods in the Advance of Cycle Studies," *Geographical Review*, XIII (October, 1923), 674–76.

______."Historical Address," *The Inaugural Bulletin* (University of Arizona, 1923), 54–59.

______."An Hypothesis Regarding the Surface Markings of Jupiter," *Popular Astronomy*, VIII (November, 1900), 473–75.

______."Illusions of Vision and the Canals of Mars," *Popular Science Monthly*, LXX (May, 1907), 464–74.

______."The Lick Review of 'Mars'," *Popular Astronomy*, IV (October, 1896), 199–201.

______."The Lowell Observatory and Its Work," *Popular Astronomy*, II (May, 1895), 395–402.

______."The Lowell Observatory in Mexico," *Popular Astronomy*, IV (March, 1897), 489–95.

______."The Markings on Venus," *Monthly Notices of the Royal Astronomical Society*, LVIII (May 13, 1898), 382–85.

______."Mars," *Popular Astronomy*, VII (March, 1899), 113–17.

______."A Method of Estimating Rainfall by the Growth of Trees," in Ellsworth Huntington, *The Climatic Factor* (Washington: Carnegie Institution of Washington, 1914), 101–22.

______."Notes on the Technique of Tree-Ring Analysis, I," *Tree-Ring Bulletin*, VII (July, 1940), 2–8.

______."Notes on the Technique of Tree-Ring Analysis, II," *Tree-Ring Bulletin*, VII (April, 1941), 28–34.

______."Notes on the Technique of Tree-Ring Analysis. III," *Tree-Ring Bulletin*, VIII (October, 1941), 10–16.

______."Notes on the Technique of Tree-Ring Analysis. IV: Practical Instruments," *Tree-Ring Bulletin*, X (July, 1943), 2–8.

______."Notes on the Technique of Tree-Ring Analysis. V: Practical Instruments," *Tree-Ring Bulletin*, X (October, 1943), 10–16.

______."An Optical Periodograph," *Astrophysical Journal*, XLI (April, 1915), 173–86.

______."A Photographic Periodogram of the Sunspot Numbers," *Astrophysical Journal*, XL (October, 1914), 326–31.

______."Present Rotation-Period of the First Satellite of Jupiter and Its Change in Form and Period Since 1892," *Astronomical Journal*, XIX (August 1, 1898), 69–70.

______."Projections on the Terminator of Mars and Martian Meteorology," *Astronomische Nachrichten*, CXLII (1897), 363–66.

______."The Secret of the Southwest Solved by Talkative Tree Rings," *The National Geographic Magazine*, LVI (December, 1929), 736–70.

______."Solar Records in Tree Growth," *Science*, March 4, 1927, 220–21.

______."Some Aspects of the Use of the Annual Rings of Trees in Climatic Study," *Annual Report. . . of the Smithsonian Institution. . . 1922* (Washington: Government Printing Office, 1924), 223–39.

______."Some Aspects of the Use of the Annual Rings of Trees in Climatic Study," *The Scientific Monthly*, XV (July, 1922), 5–21.

______."Some Topographic and Climatic Characters in the Annual Rings of the Yellow Pines and Sequoias of the Southwest," American Philosophical Society, *Proceedings*, LXI (1922), 117–22.

______."Southwestern Dated Ruins: V," *Tree-Ring Bulletin*, V (1938), 10–13.

______."The Steward Observatory of the University of Arizona," Astronomical Society of the Pacific, *Publications*, XXX (October, 1918), 326–30.

______."The Study of Atmospheric Currents by the Aid of Large Telescopes and the Effect of Such Currents on the Quality of Seeing," *American Meteorological Journal*, XI (March, 1895), 395–413.

______."A Summary of Planetary Work at the Lowell Observatory and the Conditions Under Which It Has Been Performed," *Popular Astronomy*, VII (February, 1899), 74–85.

______."Tree Rings and Climate," *The Scientific Monthly,* XXI (July, 1925), 95–99.

______."Tree Rings and Their Relation to Solar Variations and Chronology," *Annual Report . . . of the Smithsonian Institution . . . 1931* (Washington: Government Printing Office, 1932), 304–12.

______."Typical Ring-Record from Chaco Canyon, 700 to 850, CK-331," *Tree-Ring Bulletin,* III (July, 1937), 20–21.

______."The University of Arizona Eclipse Expedition, Port Libertad, Sonora, Mexico, September 10, 1923," Astronomical Society of the Pacific, *Publications,* XXXVI (August, 1924), 170–84.

______."Weather Cycles in the Growth of Big Trees," *Monthly Weather Review,* XXXVII (June, 1909), 225–37.

______. "Zodiacal Light and Counterglow and the Photography of Large Areas and Faint Contrast," *Photographic Journal,* LVI (February, 1916), 44–48.

Eddy, John A. "The Maunder Minimum," *Science,* June 18, 1976, 1189–1202.

"Edmund Schulman, 1908–1958," *Tree-Ring Bulletin,* XXII (December, 1958), 2–6.

Evans, J.E., and E.W. Maunder. "Experiments as to the Actuality of the 'Canals' Observed on Mars," *Monthly Notices of the Royal Astronomical Society,* LXIII (June, 1903), 488–99.

Fritz, Emanuel. "Problems in Dating Rings of California Coast Redwood," *Tree-Ring Bulletin,* VI (January, 1940), 19–21.

Gibson, Charles D. "On Gladwin's Methods of Correlation in Tree-Ring Analysis," *American Anthropologist,* XLIX (April–June, 1947), 337–40.

Gladwin, Harold S. "Methods and Instruments for Use in Measuring Tree-Rings." *Medallion Papers, No. 27.* Globe: Gila Pueblo, 1940.

______."Tree-Ring Analysis: Methods of Correlations," *Medallion Papers, No. 28.* Globe: Gila Pueblo, 1940.

______."A Review and Analysis of the Flagstaff Culture." *Medallion Papers, No. 31.* Globe: Gila Pueblo, 1943.

______."Tree-Ring Analysis, Problems of Dating. I: The Medicine Valley Sites." *Medallion Papers, No. 32.* Globe: Gila Pueblo, 1944.

______."Tree-Ring Analysis: Problems of Dating. II: The Tusayan Ruin." *Medallion Papers, 36.* Globe: Gila Pueblo, 1946.

______."Tree-Ring Analysis: Tree-Rings and Droughts." *Medallion Papers, 37.* Globe: Gila Pueblo, 1947.

Harris, J. Arthur. "The Correlation Between Sun-Spot Number and Tree Growth," *Monthly Weather Review*, LIV (January, 1926), 13–14.

Hartmann, William K. *Moons and Planets: An Introduction to Planetary Science*. Belmont, California: Wadsworth Publishing Company, Inc., 1972.

Haury, Emil W. "Climate and Human History," *Tree-Ring Bulletin*, I (October, 1934), 13–15.

______."Editorial," *Tree-Ring Bulletin*, IV (October, 1937), 2.

______."Southwestern Dated Ruins: II," *Tree-Ring Bulletin*, IV (January, 1938), 3–4.

______."HH-39: Recollections of a Dramatic Moment in Southwestern Archaeology," *Tree-Ring Bulletin*, XXIV (May, 1962), 11–14.

Haury, Emil W., and L.L. Hargrave. "Recently Dated Pueblo Ruins in Arizona," *Smithsonian Miscellaneous Collections*, LXXXII, No. 11 (August 18, 1931).

Hawley, Florence M. "The Significance of the Dated Prehistory of Chetro Ketl, Chaco Canyon, New Mexico." University of New Mexico *Bulletin 246*. Albuquerque: University of New Mexico, 1934.

Henry, A.J. "Clements on Drought Periods and Climatic Cycles," *Monthly Weather Review*, L (March, 1922), 127–31.

______."Douglass on Climatic Cycles and Tree-Growth," *Monthly Weather Review*, L (March, 1922), 125–27.

Holden, Edward S. "Addendum—The Arequipa Observations of *Mars* and *Jupiter* in 1892," Astronomical Society of the Pacific, *Publications*, VI (August 15, 1894), 225–27.

______."The Lowell Observatory, in Arizona," Astronomical Society of the Pacific, *Publications*, VI (June 9, 1894), 160–69.

Hoyle, Fred. *Astronomy and Cosmology: A Modern Course*. San Francisco: W.H. Freeman and Company, 1975.

Hoyt, William Graves. *Lowell and Mars*. Tucson: The University of Arizona Press, 1976.

Huntington, Ellsworth, "The Secret of the Big Trees," *Harper's Monthly Magazine*, CXXV (July, 1912), 292–302.

______."The Solar Hypothesis of Climatic Changes," Geological Society of America, *Bulletin*, XXV (November 2, 1914), 477–590.

Jastrow, Robert, and Malcolm H. Thompson. *Astronomy: Fundamentals and Frontiers*. Third edition. New York: John Wiley & Sons, 1977.

Judd, Neil M. "The Pueblo Bonito Expedition of the National Geographic Society," *The National Geographic Magazine*, XLI (March, 1922), 323–31.

______."Pueblo Bonito, The Ancient," *The National Geographic Magazine*, XLIV (July, 1923), 99–108.

Keenan, Philip C. "The Measurement of Lunar Color Differences," Astronomical Society of the Pacific, *Publications*, XLIII (June, 1931), 203–14.

Kidder, A.V. "Southwestern Archaeological Conference," *Science*, November 18, 1927, 489–91.

Langley, Samuel P. "On a Possible Variation of the Solar Radiation and its Probable Effect on Terrestrial Temperatures," *Astrophysical Journal*, XIX (June, 1904), 305–21.

Lockyer, Norman, and W.J.S. Lockyer. "On Solar Changes of Temperature and Variations in Rainfall in the Region Surrounding the Indian Ocean," *Proceedings of the Royal Society of London*, LXVII (1901), 409–31.

Lowell, A. Lawrence. *Bibliography of Percival Lowell*. New York: The Macmillan Company, 1935.

Lowell, Percival. "Explanation of the Supposed Signals from Mars of December 7 and 8, 1900," American Philosophical Society, *Proceedings*, XL (December, 1901), 166–76.

______. *Mars*. Boston: Houghton, Mifflin & Co., 1895.

______."On the Existence of a Twilight Arc upon the Planet Mars," *Astrophysical Journal*, II (August, 1895), 136–47.

Martin, Douglas D. *The Lamp in the Desert: The Story of the University of Arizona*. Tucson: The University of Arizona Press, 1960.

McGregor, John C. *Southwestern Archaeology*. Second edition. Urbana: University of Illinois Press, 1965.

Newcomb, Simon. "A Search for Fluctuations in the Sun's Thermal Radiations through their Influence on Terrestrial Temperature," American Philosophical Society, *Transactions*, new series, XXI (1908), 309–87.

Nordmann, Charles. "The Periodicity of Sun Spots and the Variations of the Mean Annual Temperatures of the Atmosphere," *Monthly Weather Review*, XXXI (August, 1903), 371.

Overstreet, Daphne. "The Man Who Told Time by the Trees," *The American West*, XI (September, 1974), 26–29, 60–61.

Pickering, William H. "The Mountain Station of the Harvard College Observatory," *Astronomy and Astro-Physics*, V (May, 1892), 353–57.

______."The Arequipa Observations of *Mars* and *Jupiter*," Astronomical Society of the Pacific, *Publications*, VI (August 15, 1894), 221–25.

______."Artificial Disks," *Annals of the Astronomical Observatory of Harvard College*, XXXII, part II (1900), 117–57.

______."The Planets," *Annals of the Astronomical Observatory of Harvard College*, XXXII, part II (1900), 158–73.

______."Brighter Satellites of Jupiter and Saturn," *Annals of the Astronomical Observatory of Harvard College*, LXI, part I (1908), 72–87.

Pinkayan, Subin. "Areal Distribution of Wet and Dry Years." Ph.D. dissertation, Colorado State University, 1965.

Robinson, William J. "Tree-Ring Dating and Archaeology in the American Southwest," *Tree-Ring Bulletin*, XXXVI (1976), 9–20.

______."Tree-Ring Materials as a Basis for Cultural Interpretations." Ph.D. dissertation, University of Arizona, 1967.

Schaeberle, J.M. "Preliminary Note on the Observations of the Surface Features of *Mars* during the Opposition of 1892," Astronomical Society of the Pacific, *Publications*, IV (September 3, 1892), 196–98.

"Schiaparelli's Observations of Mars," *The Observatory, A Monthly Review of Astronomy*, V (May 1, 1882), 138–43.

Schulman, Edmund. "Variations between Ring Chronologies in and near the Colorado River Drainage Area," *Tree-Ring Bulletin*, VIII (April, 1942), 26–32.

______."Bristlecone Pine, Oldest Living Thing," *The National Geographic Magazine*, CXIII (March, 1958), 355–72.

Schuster, Arthur. "On the Investigation of Hidden Periodicities with Application to a Supposed 26 Day Period of Meteorological Phenomena," *Terrestrial Magnetism*, III (March, 1898), 13–41.

______."The Periodogram of Magnetic Declination as obtained from the records of the Greenwich Observatory during the years 1871–1895," Cambridge Philosophical Society, *Transactions*, XVIII (1900), 107–35.

See, T.J.J. "A Sketch of the New 24-inch Refractor of the Lowell Observatory," *Popular Astronomy*, IV (December, 1896), 297–300.

Senter, Florence H. "Dendrochronology: Can We Fix Prehistoric Dates in the Middle West by Tree Rings?" *Indiana History Bulletin*, XV (February, 1938), 118–27.

Stallings, W.S., Jr. "A Tree-Ring Chronology for the Rio Grande Drainage in Northern New Mexico," National Academy of Sciences, *Proceedings*, XIX (September 15, 1933), 803–806.

Stewart, Milroy N. "Relation of Precipitation to Tree Growth," *Monthly Weather Review*, XLI (September, 1913), 1287.

Studhalter, R.A. "Early History of Crossdating," *Tree-Ring Bulletin*, XXI (April, 1956), 31–35.

Sykes, Glenton G. "A Sonoran Adventure: The University of Arizona's Eclipse Expedition of September, 1923," *Mercury*, I (January/February, 1972), 12–17.

"Symposium on Climatic Cycles," National Academy of Sciences, *Proceedings*, XIX (1933), 349–88.

Webb, George E. "The Indefatigable Astronomer: A.E. Douglass and the Founding of the Steward Observatory," *The Journal of Arizona History*, XIX (Summer, 1978), 169–188.

______. "A.E. Douglass and the Canals of Mars," *The Astronomy Quarterly*, III, number 9 (1979), 27–37.

Wells, H.G. *The War of the Worlds*. London: William Heinemann, 1898.

Whitney, Charles A. *The Discovery of Our Galaxy*. New York: Alfred A. Knopf, 1971.

Index

Abbot, Charles Greely, 102–103, 157, 169
Adams, Walter S., 75
Aitken, R. G., 75, 76
Allantown (Arizona), 150
Altadena (California), 51, 181
American Academy of Arts and Sciences, 177
American Anthropological Association, 129
American Association for the Advancement of Science, 41, 109, 119; Southwestern Division, 70, 91, 92
American Geographical Society, 177
American Museum of Natural History, 114, 116, 117
American Philosophical Society, 123, 177
Archaeology, 114–19, 131–51. *See also* Dendrochronology, names of specific sites
Arequipa (Peru), 8, 9, 12, 25, 31–33, 35, 46
Arizona Board of Regents, 60, 67 84, 87–90, 97, 158
Arizona State Museum, 160
Arizona, University of, 17, 52, 54, 57–58, 65, 68, 81–84, 86, 87, 89–93, 100, 104–105, 109, 125–27, 129, 136–137, 147, 149, 155, 158–61 163, 167, 169, 175, 177, 186; astronomy at, 55–58. *See also* Steward Observatory; Laboratory of Tree-Ring Research
Astroclimatology, 178–85, 188
Astronomical Society of the Pacific, 27, 79
Astronomy, 4–50, 54–81, 95–100,

(Astronomy cont.)
172, 173. *See also* Observatories; names of individual planets
Aztec (New Mexico) 114–16, 118, 144

Babcock, Kendrick C., 57, 82, 83 86, 90
Baer, John M., 85
Bailey, Solon I., 8
Baker Ranch (Arizona), 150
Bancroft, Hubert H., 110
Bandelier, Adolph F., 115, 141
Bannister, Bryant, 186, 188
Barnard, E. E., 39–41, 46
Beal, R. S., 93, 94
Betatakin (Arizona), 144
Boothroyd, Samuel, 38, 42
Boulder Dam, 175, 176
Bovard, G. F., 85
Boyden, Uriah A., 7
Brashear Company, 22, 58, 63, 65–68, 71, 74
Brashear, John A., 21
Bristlecone Pine, 187
Brown, Ernest W., 179, 180
Burgess, Paul S., 158
Bush, Vannevar, 184

California Institute of Technology, 88, 129, 186
California, University of, 52, 77, 84
Callisto (Jovian moon), 10, 28, 35
Campbell, W. W., 28, 43, 46, 50, 59, 74, 85
Canals, Martian (*canali*), 10, 14, 23. *See also* Lowell, Percival; Mars; Schiaparelli, Giovanni
Cannon, Annie J., 9
Canyon de Chelly (Arizona), 115, 132–34, 136
Carnegie Desert Laboratory, 77, 91
Carnegie Institution, 63, 88, 112, 126, 129, 132, 153, 155–60, 177, 184, 185
Carpenter, Edwin F., 77, 80, 81 97, 158, 172, 173
Casa Blanca (Arizona), 133
Casa Grande (Arizona), 115
Cepheid variables, 80
Chaco Canyon (New Mexico), 117–19, 132, 133, 144, 149
Chetro Ketl (New Mexico), 119, 147, 149
Chicago, University of, 33, 38, 147
Chinle (Arizona), 133
Citadel (Arizona), 134, 138
Clark, Alvan G., 21, 33
Clements, Frederic E., 126
Climate, relation of solar activity to, 103, 106, 107, 180
Climatic cycles, 105, 108
Climatic Cycles and Tree-Growth: (vol. 1), 112, 113, 122; (vol. 2), 126, 127; (vol. 3), 156, 157; (vol. 4, unpublished), 185
Cogshall, Wilbur A., 33, 36–38
Colton, Harold S., 132, 134, 137, 141, 146, 156, 160, 167–70, 185, 186, 189
Comstock, Theodore, B., 17
Cook, H. L., 72, 73
Cummings, Byron, 134, 137, 159
Cushing, Frank Hamilton, 115
Cyclograph, 152, 156, 176, 181. *See also* Periodograph

Dendrochronology: analytical methods in, 127; and Beam HH-39, 141, 143, 144, 146; bridge method in, 134–37; and Bristlecone Pine, 187; and Citadel Dating, 134, 136, 137; climate and solar-activity studies in, 102, 105, 113, 123, 125; and climatic cycles, 128, 129, 152–154, 157, 183; and climatic studies, 175–179; core samples for, 116, 121; crossdating in, 104–106,

(Dendrochronology cont.) 116, 119; and Early Pueblo Dating, 148–50; early studies in, 101, 102; and extension of chronology, 146–51; and Flagstaff record, 108, 113; and Johnson Canyon Dating, 148–50; and 179D sequence, 149; and pottery, 139–41; and redwood specimens, 121, 122, 124, 127; and Relative Dating, 116, 117, 133, 134, 136, 137, 139, 144; ruin dating through, 139, 144, 149–51; and sequoia record, 107, 109, 110, 115, 120, 123–25, 128; and solar activity–tree growth studies, 122; specimens for, 127; tree rings and climate in, 119–21
Denver (Colorado), 39
Done, Williams, 78
Douglass, Andrew Ellicott: academic honors, 3, 83, 84, 159, 160; administrative duties, 86–89, 91–92; in Arequipa (Peru), 8–10, 12; cycle concept of, 153; dismissal from Lowell Observatory, 48, 49; early dendrochronology techniques of, 102 104; early education, 2, 3; and evolution controversy, 93, 94; graduate study, 5, 6, 12; and Harvard College Observatory, 7; marriage, 51; and periodograph, 110–12; political activity of, 50, 51; publications, 29, 41–44, 58, 79, 92, 105, 112, 113, 119, 125–27, 144, 145, 147, 148, 156, 157, 169, 185, 188; as teacher, 52, 82–84, 87, 89, 154, 155; tree ring and precipitation studies of, 102, 103, 110; and University of Arizona observatory, 54–56, 58–60
Douglass, Ida Whittington, 51, 85, 86, 88, 124, 189
Douglass, Malcolm (father), 2
Douglass, Malcolm (nephew), 133
Douglass, Sarah Elizabeth (Hale), 2, 85
Drew, David A., 33, 38
Dunseath, James R., 97

Eclipses, solar, 3, 12, 44, 77, 78, 88
Eliot, Charles W., 84
Europa (Jovian moon), 28, 35

Fewkes, Jesse Walter, 115
Field Museum of Natural History, 145
Flagstaff (Arizona), 18–20, 22, 23, 25, 29–34, 36–41, 43, 48–53, 59, 82, 83, 85, 86, 89, 103–106, 113, 121, 128, 129, 132, 133, 137, 138, 140, 141, 146, 156, 160, 166–68, 175
Flammarion, Camille, 86
Forestdale (Arizona), 140
Fox, Philip, 60
Freeman, Merrill P., 56, 59, 60
Frost, Edwin B., 59, 88

Ganymede (Jovian moon), 28, 35
Gegenschein, 14
Gila Pueblo Archaeological Foundation, 161, 163, 186
Gladwin, Harold S., 161–71, 186
Globe (Arizona), 161
Glock, Waldo S., 156
Grosvenor, Gilbert, 132

Hale, George Ellery, 38, 45, 59, 64, 74, 85
Halley's Comet, 58, 59
Hargrave, Lyndon L., 137–41, 143, 144
Harris, J. Arthur, 125
Harvard College Observatory, 6, 7, 13, 14, 21, 31, 32, 44, 50,

(Harvard Observatory cont.) 51, 57, 83, 84, 98, 111; Boyden Station of, 8–11, 43
Harvard University, 12, 13, 41, 77, 84, 88, 166
Harvill, Richard A., 187
Haury, Emil W., 140, 141, 143, 144, 158, 161, 162, 167
Hawaiian Islands, 32, 39
Hawikuh (New Mexico), 131
Hawley, Florence M., 147, 149, 155
Hewett, Edgar L., 91, 115, 134
Heye, George Gustav, 117
Hodge, Frederick Webb, 117
Hohokam Indians, 161
Holden, Edward S., 27
Hopi Indians, 132, 137, 138
Hubble, Edwin P., 80
Hughes, John T., 59
Hungo Pavi (New Mexico), 119
Hunt, George W. P., 55, 60
Huntington, Ellsworth, 106–108, 114

Io (Jovian moon), 28, 35, 39

Jastrow, Joseph, 48
Jeddito (Arizona), 139
Johns Hopkins University, 5, 12
Johnson Canyon (Colarado), 148, 149
Jones, Inigo, 180
Judd, Neil M., 117, 118, 129, 131–34, 138, 140, 141, 143–45, 147, 148
Jupiter, 4, 9, 10, 15, 28, 35, 39, 42, 44, 58, 179, 181; moons of, 10, 28, 35, 39

Kawaiku (Arizona), 139
Kayenta (Arizona), 133
Keet Seel (Arizona), 144
Kidder, Alfred V., 115, 134
Kinbiniyol (New Mexico), 133
Kin Tiel (Arizona), 140, 144
KleinSmid, R. B. von, 60, 61, 67, 73, 89, 92, 95
Kokopnyama (Arizona), 139, 140, 144
Kroeber, Alfred L., 134, 145

Laboratory of Anthropology (Sante Fe, New Mexico), 147, 186
Laboratory of Tree-Ring Research, 160, 175–78, 185–90; early plans for, 129; established, 158
Langley, Samuel P., 102
Lick Observatory, 25, 27–29, 38, 39, 43, 46, 50, 61, 74, 75, 84, 85, 98, 99
Lockyer, Sir Norman, 86
Lockyer, W. J. S., 86
Lowell Observatory, 14–19, 20–23, 26, 27, 29, 30, 31–36, 32, 38, 39, 41, 43, 47–49, 58, 64, 69, 75, 80, 82, 83, 89, 91
Lowell, Percival, 13–28, 30–41, 43–51, 59
Luther, Flavel S., 4, 5, 52, 84, 90

McCormick, J. Byron, 186
MacDougal, D. T., 91, 126
McDowell, James B., 63, 66–68, 71, 74
McGinnies, William G., 188
McGregor, John C., 146, 150, 165, 166, 168
Mars, 4, 5, 9, 10, 12–14, 19, 20, 23–27, 29–31, 35, 39, 42–46, 48, 51, 58, 64, 65, 74, 79, 80, 92, 183
Mars Hill, 21, 29, 30, 33, 37
Martin, Paul S., 145
Martindale, Roy, 175, 176
Martz, E. P., 80
Marvin, Cloyd H., 75, 77, 78, 93, 125
Maunder, E. Walter, 122, 123
Medallion Papers, 163, 164, 166
Medicine Valley (Arizona), 166, 168
Mercury, 9, 10, 39, 183
Merriam, John C., 126, 129, 155–57, 159

Mesa Verde (Colorado), 114, 115, 132, 134, 144, 148, 150
Mexico City, 31, 32
Milky Way Galaxy, 80
Morris, Earl H., 115, 116, 136, 139, 140, 149
Mummy Cave (Arizona), 134, 148
Museum of Northern Arizona, 137, 140, 146, 150, 160, 165, 167, 185, 186
Museum of the American Indian, 117

National Academy of Sciences, 152
National Geographic Society, 117, 129, 131, 132, 136, 137–39, 140, 143–44, 147, 154, 177
National Science Foundation, 187
Navajo Indians, 133
Nebulae, spiral, 9, 80
Neptune, 9, 10, 83
Newcomb, Simon, 103, 106
New Mexico, University of, 147
Northern Arizona Normal School, 52, 82, 83, 85, 86

Obelisk Cave (New Mexico), 150
Observatories: Greenwich (England), 13; Griffith (Los Angeles, California), 80; Kitt Peak National (Arizona), 173; Mexican National (Tacubaya), 32; Mt. Lowe (Altadena, California), 51; Mt. Palomar (California), 64; Mt. Wilson (California), 63, 64, 75, 80, 85, 98, 129; Smithsonian Astrophysical (Washington, D.C.), 102; Solar Physics (England), 86; U.S. Naval (Washington, D.C.), 38, 50, 77; Yerkes (Wisconsin), 39, 59, 88, 98. *See also* Harvard College Observatory; Lick Observatory; Lowell Observatory; Steward Observatory
Occidental College, 85, 87
Oracle (Arizona), 16, 61, 173
Oraibi (Arizona), 133, 137, 138, 139, 144
Ott, H. N., 68, 70

Palo Alto (California), 129
Pasadena (California), 63, 129
Peabody Museum of American Anthropology and Ethnology, 12, 166
Pecos classification, 135
Pecos Conference, 134, 135
Pecos (New Mexico), 131, 134
Peñasco Blanco (New Mexico), 117, 119
Peru, 7, 8, 9, 12
Petrified Forest (Arizona), 133
Phoenix (Arizona), 14, 16–18, 92, 173, 177
Photography, 5, 9, 12, 20, 58, 74, 78–80
Pickering, Edward C., 6, 9, 22, 31, 57
Pickering, William H., 6–8, 10, 13–15, 20, 21, 23, 25, 28, 30, 32, 35, 41, 44, 46, 48, 56, 57, 59
Pinedale (Arizona), 140, 144
Pluto, 79
Prentiss, Elliot C., 91
Prescott (Arizona), 17–19, 105, 106, 110, 120
Pueblo Bonito (New Mexico), 115–19, 131–34, 136, 139, 140, 144, 145, 147
Pueblo del Arroyo (New Mexico), 119, 131
Puerto Libertad (Mexico), 77, 78
Putnam, F. W., 12, 49, 116
Putnam, William L., 37, 38, 40, 41

Riordan, T. A., 103
Rockefeller Foundation, 159, 160
Ruin J (Arizona), 134

Russell, Henry Norris, 153

Sage, Russell (Foundation), 159
Sahara Desert, 33
Santa Fe (New Mexico), 18, 39, 91, 134, 147, 161
Saturn, 13, 41, 44, 179; moons of, 41, 44
Schaeberle, J. M., 25
Schiaparelli, Giovanni, 10, 14, 23, 25, 46
School of American Archaeology, 91
Schulman, Edmund, 156, 165–67, 175–78, 186, 187
Schuster, Arthur, 111
Scripps Institution of Oceanography, 175
See, Thomas J. J., 33, 36–38
Seeliger, Hugo von, 88
Seely, Warner, 67, 156
Shantz, H. L., 97, 98, 155, 156
Shapley, Harlow, 98, 99
Sharp, Donald E., 74
Showalter, William J., 145, 154
Show Low (Arizona), 140, 141, 143–46
Sliding Ruin (Arizona), 136
Slipher, Vesto M., 75, 80, 91
Sloan, Richard A., 87
Smiley, Terah L., 178, 186–88
Smithsonian Institution, 117, 150, 157
Solomon (New Mexico), 133
Southern California, University of, 85, 87
Spencer Lens Company, 67, 68 70, 71, 73, 74
Spoehr, H. A., 153, 156
Stallings, Williams S., Jr., 146, 147, 155, 186
Stanford University, 129
Stauffer, L. B., 66
Step House (Colorado), 150
Steward, Lavinia, 61, 65, 66, 173; bequest of, 61
Steward Observatory, 54, 61, 63–76, 79–81, 92, 95, 97–100, 123 126, 156, 157, 172–73
Stokes, Marvin A., 186
Sunspots, 103, 105–107, 113, 122, 123, 128, 179–84, 188
Sykes, F. H., 90
Sykes, Glenton, 78
Sykes, Godfrey, 34, 37, 69, 77

Tacubaya (Mexico), 32–34
Taylor, A. N., 85
Tempe (Arizona), 17, 18
Tombstone (Arizona), 14–17
Tree-Ring Bulletin, 160–62
Tree Ring Conference, 160, 161
Tree Ring Society, 161
Tree rings. *See* Dendrochronology
Trinity College, 2–5, 52, 83, 84
Tucson (Arizona), 14, 16, 17, 21, 22, 52–57, 59–61, 65, 68–72, 74, 77, 79–83, 85, 86, 88, 91, 93, 95, 99, 109, 114, 117, 118, 121, 132, 139, 157–59, 175, 185; city council of, 95, 96

Uranus, 183

Valle, Felipe, 33
Venus, 9, 10, 39, 43, 44, 47, 48, 74, 183

Wallace, Henry A., 177
Walpi (Arizona), 138
Warner and Swasey Company, 27, 61, 64–67, 70–73
Warner, W. R., 66
Wetherill, John, 133
Whipple (Arizona), 140, 141
Whipple, Fred L., 184
Wilde, Arthur H., 56, 87
Wissler, Clark, 114, 116
Wright, Boardman, 90
Wright, W. H., 99
Wupatki (Arizona), 132–34, 144

Yuma (Arizona), 17

Zodiacal light, 12, 58